WORKSHOPS IN COMPUTING
Series edited by C. J. van Rijsbergen

Also in this series

Women into Computing: Selected Papers 1988–1990
Gillian Lovegrove and Barbara Segal (Eds.)

3rd Refinement Workshop (organised by
BCS-FACS, and sponsored by IBM UK
Laboratories, Hursley Park and the Programming
Research Group, University of Oxford),
Hursley Park, 9–11 January 1990
Carroll Morgan and J. C. P. Woodcock (Eds.)

Designing Correct Circuits, Workshop jointly
organised by the Universities of Oxford and
Glasgow, Oxford, 26–28 September 1990
Geraint Jones and Mary Sheeran (Eds.)

Functional Programming, Glasgow 1990,
Proceedings of the 1990 Glasgow Workshop on
Functional Programming, Ullapool, Scotland,
13–15 August 1990
Simon L. Peyton Jones, Graham Hutton and
Carsten Kehler Holst (Eds.)

4th Refinement Workshop, Proceedings of the
4th Refinement Workshop, organised by BCS-
FACS, Cambridge, 9–11 January 1991
Joseph M. Morris and Roger C. Shaw (Eds.)

AI and Cognitive Science '90, University of
Ulster at Jordanstown, 20–21 September 1990
Michael F. McTear and Norman Creaney (Eds.)

Software Re-use, Utrecht 1989, Proceedings of
the Software Re-use Workshop, Utrecht,
The Netherlands, 23–24 November 1989
Liesbeth Dusink and Patrick Hall (Eds.)

Z User Workshop, 1990, Proceedings of the Fifth
Annual Z User Meeting, Oxford,
17–18 December 1990
J.E. Nicholls (Ed.)

IV Higher Order Workshop, Banff 1990
Proceedings of the IV Higher Order Workshop,
Banff, Alberta, Canada, 10–14 September 1990
Graham Birtwistle (Ed.)

ALPUK 91 Proceedings of the 3rd UK
Annual Conference on Logic Programming,
Edinburgh, 10–12 April 1991
Geraint A. Wiggins, Chris Mellish and
Tim Duncan (Eds.)

Specifications of Database Systems,
International Workshop on Specifications of
Database Systems, Glasgow, 3–5 July 1991
David J. Harper and Moira C. Norrie (Eds.)

continued on back page...

J. Hillston, P.J.B. King and R.J. Pooley (Eds.)

7th UK Computer and Telecommunications Performance Engineering Workshop

Edinburgh, 22–23 July 1991

Springer-Verlag London Ltd.

ISBN 978-3-540-19733-1   ISBN 978-1-4471-3538-8 (eBook)
DOI 10.1007/978-1-4471-3538-8

British Library Cataloguing in Publication Data
UK Computer and Telecommunications Performance Engineering Workshop (7th:
1991: Edinburgh, Scotland)
7th UK Computer and Telecommunications Performance Engineering Workshop.
– (Workshops in computing)
I. Hillston, J.   II. King, P.J.B.   III. Pooley, R.J.   IV. Series
004

Library of Congress Data available

34/3830-543210 Printed on acid-free paper

Preface

This book contains the papers presented at the UK Computer and Telecommunications Performance Engineering Workshop in Edinburgh in July 1991. The Workshop series has been running annually since 1985, providing a valuable forum for academic and industrial practitioners to meet and exchange experiences. Previous workshops have been held in Edinburgh and Bradford under the auspices of the Performance Engineering Group of the British Computer Society and with support from UKCMG. The scope of the papers ranges from theoretical advances in queueing theory to practical papers expounding common performance problems and their solutions. It is hoped that the publication of the proceedings will encourage more people to apply their talents to this challenging field, where advances in technology provide new problems regularly.

In the interests of a coherent format to all papers in the proceedings, a number of papers were converted to LaTeX from other word processor formats, and proof read by the editors. The editors would like to thank all the contributors and participants to this and previous workshops. Particular thanks are due to Margaret Davies of Edinburgh University for her calm efficiency organising the Workshop and Stephen Gilmore for his kind help with the preparation of these Proceedings.

Edinburgh
August 1991

Jane Hillston
Peter King
Rob Pooley

Contents

List of Contributors ix

The Performance Analysis Process
Rob Pooley and Jane Hillston 1

A Statistical Approach to Finding Performance Models of Parallel
Programs
Rosemary Candlin, Peter Fisk and Neil Skilling 15

A Modelling Environment for Studying the Performance of
Parallel Programs
Joe Phillips and Neil Skilling 27

On the Performance Prediction of PRAM Simulating Models
George Chochia 40

General Queueing Network Models with Variable Concurrent
Programming Structures and Synchronisation
Demetres D. Kouvatsos and Andreas Skliros 56

Experiences in Implementing Solution Techniques for Networks
of Queues
David Ll. Thomas 73

BCMP Queueing Networks versus Stochastic Petri Nets:
A Pragmatic Comparison
Peter J.B. King and Saqer Abdel-Rahim 88

Extension to Preemption of a Method for a Feedback Queue
Dick H.J. Epema 99

Approximate Analysis of a G/G/c/PR Queue
Nasreddine M. Tabet-Aouel and Demetres D. Kouvatsos 108

The CLOWN Network Simulator
Soren-Aksel Sorensen and Mark G.W. Jones 123

Modelling ATM Network Components with the Process
Interaction Tool
Nick Xenios and Peter Hughes 131

Asynchronous Packet-switched Banyan Networks with Blocking
(Extended Abstract)
Peter G. Harrison and Afonso de C. Pinto 146

Equilibrium Point Analysis of a Slotted Ring
Michael E. Woodward 150

MEM for Arbitrary Exponential Open Networks with Blocking
and Multiple Job Classes
Demetres D. Kouvatsos, Spiros G. Denazis and
Panagiotis H. Georgatsos 163

Performance Evaluation of FDDI by Emulation
Frank Ball and David Hutchison 179

An Editing and Checking Tool for Stochastic Petri Nets
Auyong Lin Song 185

Author Index 189

List of Contributors

Saqer Abdel-Rahim
Department of Computer Science, Heriot-Watt University,
79 Grassmarket, Edinburgh, EH1 2HJ

Frank Ball
Computing Department, Lancaster University, Engineering Building,
Lancaster, LA1 4YR

Rosemary Candlin
Department of Computer Science, James Clerk Maxwell Building,
Edinburgh University, The King's Buildings, Mayfield Road,
Edinburgh, EH9 3JZ

George Chochia
Edinburgh Parallel Computing Centre, James Clerk Maxwell Building,
Edinburgh University, The King's Buildings, Mayfield Road,
Edinburgh, EH9 3JZ

Spiros G. Denazis
Computer Systems Modelling Research Group, Computer Science,
University of Bradford, Bradford, BD7 1DP

Dick H.J. Epema
Department of Mathematics and Computer Science, Delft University of
Technology, P.O. Box 356, 2600 AJ Delft, The Netherlands

Peter Fisk
Department of Statistics, James Clerk Maxwell Building,
Edinburgh University, The King's Buildings, Mayfield Road,
Edinburgh, EH9 3JZ

Panagiotis H. Georgatsos
Telecommunications Division, ALFA SAI, 72-74 Salaminos Street,
Kallithea, Athens, Greece

Peter G. Harrison
Department of Computing, Imperial College, 180 Queen's Gate,
London SW7 2BZ

Jane Hillston
Department of Computer Science, James Clerk Maxwell Building,
Edinburgh University, The King's Buildings, Mayfield Road,
Edinburgh, EH9 3JZ

Peter Hughes
BNR Europe Ltd, London Road, Harlow, Essex CM17 9NA

David Hutchison
Computing Department, Lancaster University, Engineering Building,
Lancaster LA1 4YR

Mark G.W. Jones
Department of Computer Science, University College London,
London WC1E 6BT

Peter J.B. King
Department of Computer Science, Heriot-Watt University,
79 Grassmarket, Edinburgh, EH1 2HJ

Demetres D. Kouvatsos
Computer Systems Modelling Research Group, Computer Science,
University of Bradford, Bradford, BD7 1DP

Joe Phillips
Department of Computer Science, James Clerk Maxwell Building,
Edinburgh University, The King's Buildings, Mayfield Road,
Edinburgh, EH9 3JZ

Afonso de C. Pinto
Department of Computing, Imperial College, 180 Queen's Gate,
London SW7 2BZ

Rob Pooley
Department of Computer Science, James Clerk Maxwell Building,
Edinburgh University, The King's Buildings, Mayfield Road,
Edinburgh, EH9 3JZ

Neil Skilling
Department of Chemical Engineering, James Clerk Maxwell Building,
Edinburgh University, The King's Buildings, Mayfield Road,
Edinburgh, EH9 3JL

Andreas Skliros
Computer Systems Modelling Research Group, Computer Science,
University of Bradford, Bradford, BD7 1DP

Auyong Lin Song
Department of Computer Science, James Clerk Maxwell Building,
Edinburgh University, The King's Buildings, Mayfield Road,
Edinburgh, EH9 3JZ

Soren-Aksel Sorensen
Department of Computer Science, University College London,
London WC1E 6BT

Nasreddine M. Tabet-Aouel
Computer Systems Modelling Research Group, Computer Science,
University of Bradford, Bradford, BD7 1DP

David Ll. Thomas
British Telecom Computer Performance Management Group,
Room 460B, Brunel House, 2 Fitzalan Road, Cardiff, CF2 1YA

Michael E. Woodward
Department of Electronic and Electrical Engineering, Loughborough
University of Technology, Loughborough, Leicestershire, LE11 3TU

Nick Xenios
BNR Europe Ltd, London Road, Harlow, Essex, CM17 9NA

The Performance Analysis Process

Rob Pooley and Jane Hillston
Department of Computer Science
Edinburgh University

Abstract

A perspective is offered for viewing the history, and projecting the future, of tools and environments for performance analysis. This is derived by attempting to capture performance analysis as a process and then considering how this process may be supported efficiently from the user point of view. It is argued that such a conceptual modelling approach will enable developers to minimise the load on users in making use of the wide and growing range of tools and techniques on offer.

1 Introduction

In this paper we discuss the use which people make, or would like to make, of performance analysis techniques, in the hope that more appropriate ways of supporting such use may emerge.

At present we are confronted with an apparently bewildering proliferation of tools and techniques which can loosely be grouped under the performance banner. In recent years the individual tools have been supplemented by "support" tools, such as expert system front and back ends [1], by integrated tools, such as HIT [2], and by toolsets and environments, such as Tangram [3] and IMSE [4]. How can potential users make sense of this cornucopia and how far do the current tools go towards addressing user needs? To answer these questions, we must step back from the technology and address the aims of their users.

The first section below introduces the idea that there are stages in performance analysis. The term "analysis" is chosen deliberately to distinguish the overall *process* involving performance techniques from its constituent *stage*, "modelling". The objective of this process is assumed to be performance *prediction*. It is conjectured that there may be several ways of viewing this process, depending on the type of user involved, and that, therefore, any support should accomodate as wide a range of these as possible or should consciously be restricted to one type of use.

In section 3 we consider the range of tools and techniques currently available and map these into the model presented. Gaps and inconsistencies are considered and questions of generality dealt with. Trends in the evolution of these tools lead us to the issue of integration within toolsets and environments, which is the subject of the following section. Consideration of IMSE and of HIT in the light of the criteria identified in the earlier sections then allows us to identify possible ways forward and questions which remain to be answered before progress can be made.

Clearly, it would be foolish to suggest that any single approach can be chosen as the "correct" one. Rather, the hope is to stimulate a diversity of approaches, but also to present a clearer view of the objectives which need to be addressed.

2   A model of the process

A user may approach the performance analysis task differently depending upon his objective. It is important to understand these objectives if we are to offer the user a technology which is suitable to his needs. Some typical objectives are described below. In practice a single user is unlikely to encompass all such objectives so we are really drawing a series of overlapping profiles.

2.1   System sizing

One common use of performance prediction is in the sizing of computer systems (also known as "dimensioning" in teletraffic circles). This is an attempt to discover the configuration of hardware and, possibly software, best suited to a customer's needs. Typically the vendor already has models and solvers in place, although these may be rather crude. In the case of small system sales there may be no systematic analysis, merely "back of the envelope" calculations. For large, popular systems, such as IBM mainframes, independent consultants may be able to offer advice and undertake modelling. Again this would probably involve the use of previously constructed models.

There is a question of the reliability of generic models when used in this way but we should note that the answers are usually not required to be very accurate as the range of options will involve quantum increases in the hardware's capabilities. However, the chief problem is probably workload capture. As the choice is often among machines of similar type the trace and monitoring information is likely to be equivalent so once captured such data will be widely applicable. Major problems centre on comparisons where the operating system or application software is significantly different on the offered systems.

Essentially, the sizer will be making the choice between different system configurations, represented by different models or by different values for parameters of generic models which define structural properties of the system modelled, e.g. the number of processors rather than their rate. He will be interested in how each of these models performs under the conditions, represented by an unvarying set of workload parameters. The internal structure of the models (and the systems which they represent) is not of primary interest; only the predicted behaviour is important.

2.2   Capacity planning

Capacity planning can be regarded as an extension of sizing; however, there are a few crucial differences.

Firstly, the primary objective is to manage existing installations rather than choose new ones. Typically the capacity planning aims to *tune* the system to a given workload and to allow it to cope with predicted increases. This requires much more detailed workload characterisation and modelling. Experimentation becomes much more important, with optimisation under constraints an important concern.

Secondly, the process is more clearly customer oriented. Indeed, tuning in some ways works counter to the interests of suppliers. Questions of balancing workload are seen as more important than looking for gains from additional hardware.

Thirdly, it can be argued that software performance becomes more important. Benchmarking of software on particular hardwares to allow applications to be run under the most efficient packages is part of the capacity planning process.

The capacity planner requires a more flexible and sophisticated technology than the sizer, but he is still constrained to characteristics of existing systems. He perceives these as fixed, with tuning and workload parameters as variables, but both constrained to ranges fixed by the system and the user community. He may only need one model but will be making decisions based on the behaviour of the model under different parameter settings.

2.3 System design

As in all engineering, computer system engineering (including software) can only approach its optimum performance and reliability if these are designed for early in the development. There have been many case studies showing the benefits of performance analysis incorporated into the design cycle in achieving performance targets for systems [5].

The design of a system involves many possibilities, using new or existing components, depending on the type of project and its objectives. The range of questions to be answered increases, as do the means available for answering them. However, performance is only one of several factors which must be considered and the designer is typically not an expert in the techniques of performance analysis. As designers are usually working at a low level, workloads are often rather abstract, end users being defined in high level terms. This can also create a problem.

This leads to the paradox that those asking the most complex questions of performance analysis are often those least able or prepared to answer them with appropriately sophisticated techniques. In comparison to the sizer and the capacity planner the designer is likely to have to consider many models as well as varying the parameters within them.

In large organisations a specialist performance group or individual may be used in a consultative rôle to help the designer address these issues. This introduces overheads of communication and inflexibility. Alternatively, design engineers often write their own models from scratch. These are usually simulation models which are very detailed, do little to support analysis and are without generality. They are frequently used once in the design process and thrown away thereafter. They almost never belong to a set of concrete design information which is carefully accumulated and maintained.

This group of users is the most challenging to performance analysis tools and recently some effort has been put into making existing technology more accessible to designers. The difficulty is to provide easy to use, flexible tools which can fit into the information and conceptual environment of such users. In the area of software Smith [5] has coined the phrase *performance engineering* to describe an integrated methodology involving software engineering and performance analysis.

4

It is interesting to note that the outcome of such an integrated approach would be models of systems which could readily be adapted as the basis for sizing and capacity planning.

2.4 Research

Performance analysis is still a very active area of research and so another style of use of the technology which must be considered is such research. Researchers want a platform on which to investigate and develop new techniques and applications of the performance analysis technology or a means of verifying results obtained from new, as yet unsupported, techniques. These users are likely to require a high degree of flexibility within the tool and facilities at a lower level than less expert users. They are less likely to be concerned with ease of use than the capabilities of the tool. Which particular aspects of performance analysis the user concentrates on will depend upon his area of research.

A typical example might be a researcher who has devised a new approximate numerical solution technique for queueing network models. He wishes to compare results from this technique with those given by existing methods in a multi-solver tool, looking at efficiency and accuracy. His need can be seen to be for a way to easily add his method to those within the tool, leaving the means for constructing such models, whether as a language or a graphical interface, unchanged.

2.5 A process model of performance analysis

Overall performance analysis can be thought of as involving four stages.

1. Workload analysis and characterisation.

2. Model construction, verification and validation.

3. Experimentation.

4. Analysis and reporting

This identification of the process of performance analysis as a whole emphasises the importance of recent developments of toolsets and environments which attempt to encompass support for all aspects of the process.

Previously tools have tended to concentrate on the detailed modelling aspects of performance, increasing the power of the solvers in these tools and adding more general means of expressing models, with little or no support provided for the other stages of the performance analysis process. In terms of the user groups identified above this approach did not really benefit anyone except possibly the researcher:

- The sizer does not want such access to the internal structure of the models and would like to have help with workload characterisation.

- The capacity planner would also probably benefit from support for workload characterisation as well as support for experimentation and processing of the results.

- The designer, like the capacity planner, needs support throughout the process. He would also benefit from tools to help him manage the models he develops, similar to, or incorporated into, the design support tools he uses.

All these aspects are generally missing from highly developed *modelling* tools.

When performance modelling is no longer viewed in isolation the necessity of software tools to manage the use of models as well as their development becomes apparent. Similarly the generation of input values for the models from real or synthesised workloads and the handling of the output from the models to produce system level results are also candidates for computer support.

Each user group focuses on certain stages of the performance analysis process more than others. This may be because they have already got information from some stage; because information is hard to get from a stage; or because, particularly for analysis and reporting, that stage is not seen as important. Particularly in the latter case this disregard may also be due to a lack of tools to support the stage in question.

3 Tools and their trends

To provide useful tools, it is best to follow two, hopefully orthogonal, paths. Firstly, one should aim to build powerful engines, such as model generators, solvers, experiment executors, data analysers. Secondly, one should aim to make this power available to its potential users in a manner which they find appropriate and easy to use. In the following sections we consider the tools available to support each of the stages identified within the performance analysis process and assess their benefits to the user groups identified.

3.1 Workload characterisation and synthesis tools

Workload analysis [6] has long been recognised as a key part of understanding computer performance. Unfortunately it is very difficult to provide general purpose tools for its support. Most operating systems allow monitoring of activities at one or more levels, but the traces generated are peculiar to the system producing them, making it necessary to provide system specific recognisers. Also the level at which such monitoring is possible is often too low level (disc accesses, I/O requests etc.) or too high level, (user commands, transaction requests) to be useful.

An attempt is made to address the format incompatibility problem in the tools Workser [7], and its successor WAT [8], produced at the Universities of Pavia and Milan. These support a library of format definitions, which define the mapping from a given operating system's trace format into the internal format of the tool. Once the trace has been read in this way it can be analysed by a single back end. These analyses can be used to identify classes of customers within a system and to generate parameters suitable for a queueing model, for example.

The problem of mapping between levels of workload is addressed in several ways, but in general these require either considerable user knowledge of the system involved or provide rather specific support.

The first approach is statistical and is based upon clustering analysis. Classes, or patterns of behaviour, are identified within a low level trace; these are then associated with higher level activities. WAT/Workser is an example of a tool with such capabilities.

In the second approach, implemented in Hughes' *Sp* tool [9], the user defines a structural model of the system in terms of a hierarchy of components. Thus a tree is formed where any internal node is a component implemented by services provided by its children and any leaf is a primitive component whose characteristics are entirely defined at that level. The user can then define mappings of units of work at a higher level onto lower level units of work in its child nodes. Such a mapping defines an implementation of the component, along with appropriate quantification. The tool can then use this to derive workloads at one level from another. There is currently little experience to evaluate this tool with.

Another recent approach [10] is to define patterns of expected trace items as regular expressions which correspond to the program being monitored. Thus the program is considered to be a finite state automaton. Generalising this approach leads to a method for parsing traces to look for higher level actions as patterns of lower level ones.

All these tools are yet to achieve widespread acceptance, and are still primarily used by their developers and other researchers. Conversely, users needing to tackle such problems are typically developing and using their own approaches. However the potential benefits from workload analysis for model parameterisation and model verification, as well as understanding the system in question, are such that with improved user interfaces and growing evidence of their usefulness these tools should gain more general acceptance. For example, a tool such as *Sp* would be particularly useful to the capacity planner who will want to translate the demands of his user community into parameters for a model developed at a much lower level.

3.2 Modelling tools

As mentioned earlier modelling tools have been the focus of considerable attention over the last decade, the tools developed growing in sophistication and scope during that period. Taking a retrospective view we can perhaps distinguish four levels of development of performance modelling tools.

Simple modelling tools such as simulation packages. These are not very interesting, offering either an analytic solver for queueing network models or a simulation solver based on some convenient simulation language. The only significant development in such tools in the last ten years is the development of graphical programming interfaces such as the Performance Analysis Workstation [11]. Although simple these tools may be sufficient for the needs of a sizer who is generally only dealing with quite crude models.

Modelling tools supporting multiple solvers usually including more than one analytic solver, one or more numerical solvers and a simulation solver. The interface to all these is a single modelling language. Again some recent tools of this sort have added graphical interfaces. QNAP2 [12] is an example of this type of tool. Such tools offer quite considerable modelling power and the addition of a graphical interface reduces the necessity for the user to fully understand the underlying engine. However the only support given is for the modelling stage of the process with no assistance for parameterising the models or interpreting their results.

Modelling tools offering an integrated methodology ; these offer a similar range of solvers to the previous category, but operate within a framework designed to be closer to system description than to performance modelling. Such tools are very recent and their use is still being explored. HIT [2], an example of a tool in this category, is described in more detail later. They represent an advance towards performance engineering, where their methodology could be combined with other aspects of software and hardware engineering approaches.

Modelling environments which offer a variety of both solution techniques and formalisms for expressing models. These form the newest of the categories and work here is only just emerging from the prototyping stage. These can also support a wider range of tools in categories other than modelling, and interaction between those tools so that they can used in a cooperating manner. Such toolsets will represent a considerable increase in performance analysis power and it seems likely that the real use of such systems, other than in teaching and research, will be as a basis for more constrained turnkey systems. Their strength for end users is probably their ability to support systematic studies through experimentation and information storage in a secure manner.

3.3 Experimentation

Experimentation has been little supported, although some tools have had control languages. The idea of experimentation with models has largely been limited to simulation models and has been visible only by the provision of variance reduction techniques in these control languages. Many support replications and some (e.g. QNAP 2) add regeneration and spectral techniques. These apply only to obtaining better estimates for single points in a parameter space, not to defining complete explorations or more complex experiments.

Some attempts have been made to use expert systems to support definition of experiments. A good example of this is provided by [13]. Since many clear rules exist for statistical aspects, it is comparatively easy to support this part of the process. Ören has defined some experimentation capabilities in his tool [14] based on Zeigler's definition of experimental frames [15], but the most complete attempt at a general, modelling tool independent experimentation facility is the IMSE Experimenter tool [16]. This is so recent that hard experience has yet to be gained of its use, but it allows definitions of parameter spaces by means of ranges, sets, constraints and dependencies and provides for an extensible range of analyses. It is intended to feed a customisable report generator.

Although sizing places few requirements for systematic experimentation, capacity planning is likely to give rise to the need to fully investigate the variable parameters within specified ranges to find the optimal performance. Similarly system design will require a flexible yet systematic approach to model investigation. Expert system style support advising the user on approaches to experimentation would be useful for all users without statistical experience, although such "knowledge" could also be implicitly built into a tool.

Another useful tool for experimentation, and analysis, can be the dynamic display of the execution of a simulation. Many papers have been published on this and most simulation tools now have some form of animation. With the emergence of a manageable number of standards for graphical interfaces, notably X windows and MicroSoft Windows, this is certain to become standard. These interfaces are also making it possible to incorporate good quality graphical display of summary results.

3.4   Analysis and Reporting

Traditionally modelling tools have provided rudimentary statistical analysis in the form of report generation. This has amounted to little more than a fixed range of options, such as means, standard deviations, throughputs etc., in a summary form. Users have sometimes been able to add dumping of raw results, but little in the way of flexible, customisable facilities has usually been offered. In particular in order to alter the results generated by a model the user has had to edit the model.

Further processing has usually been done with standard statistical packages, such as MiniTab. This can cause similar problems to workload analysis, in adapting formats to allow analysis tools to recognise results. Often this allowed generation of graphs and tables to aid readability. However, it relied upon the user manually transferring results from the modelling tool to the appropriate statistical package.

Storage of results has usually been as textual files, containing summary reports, user requested dumps of values or event traces from simulations. Little or no use is typically made of databases. Reusability of results has suffered consequently, particularly because there would be no way of knowing whether the model had changed since a set of results were generated.

Clearly all users would benefit from improved analysis and reporting services. In particular system designers would be greatly helped by any facility which could interpret the modelling results in terms of the system being represented to some extent. As yet no tool has yet emerged which has this capability and it seems essential that some form of system description must also be incorporated in any such tool, and current work is looking at the possibilities of using existing system description formalisms for model description [17].

4   Toolsets and Environments

Returning to our earlier orthogonal objectives of providing engines which are designed to be easily accessible, we have seen that at the moment many more or less powerful engines exist. Some of these have grown to include some rudimentary aspects of other engines. Such growth has usually been outwards from a model solver. The users are, as we have seen, likely to require different combinations of such engines. For many, the issue of solvers, for instance, is of little interest while for others it is crucial. We are likely to see further attempts at growing out and also the "bolting" together existing tools to form toolsets and environments.

The attraction of the bolting together approach is the wide range of facilities which can be offered to the user. Based upon established work, a large comprehensive toolkit can be built of existing tools, enhanced by new ones, with relative speed. Many different tools can be included, cooperating and sharing data in such a way that the user is offered support for the whole performance analysis process. Ideally, the whole becomes greater than the sum of its parts, the tools interacting in ways which enhance the performance analysis process and extend its applicability. In reality, there are many problems to be overcome in this approach and its success may be limited. Even the full generality offered may itself prove to be a disadvantage. The Integrated Modelling Support Environment (IMSE)[1] described below is an example of such a toolset.

The chief obstacle to the bolting together approach seems to be the reluctance of tools to revert to being their basic engines. Connecting tools each with their own version of report generation, variance reduction, multi-solver provision and selection etc. involves much perceived loss of independence and may be technically difficult if the tool is not well structured. This may limit the potential for real progress in this approach. On the other hand, throwing away the existing tools, or developing a large toolset from scratch, are also an unlikely course in the short run.

Despite the problems the benefits to be gained from a performance analysis enviroment seem attractive enough to make it worthwhile. There are several new "tools" which are needed for such an environment. These do not directly correspond to the identifed stages of performance analysis but nevertheless are essential.

Overall integration of tools requires user interface tools. At a cosmetic level, this is needed in order to make the tools appear integrated to the user. More importantly, if each tool maintained its own interface the user would be overburdened learning each one. These interface tools need to be specifically designed for performance analysis. Recent developments within modelling tools have shown that visual programming offers some attractive possibilities, and these tend to be based on similar types of attributed graph. Generic graphical editing tools for these would appear to be an essential feature of a performance toolset. For an early example see [18].

Common user interface standards may help users to switch among tools, but true integration depends on transparent sharing of data. Object oriented databases and object management systems are a powerful aid to supporting this in a secure manner. This approach has been used with some benefit in the IMSE system [19].

4.1 IMSE

The tools supported within IMSE are designed to assist in **all** stages of the performance analysis, including model construction and generation of workloads, through experimentation, to the generation of final reports on findings. The environment has been designed to support users without in-depth knowledge of the tools and techniques employed. The environment is made available to the user through a graphical interface, known as the WorkBench, and by mouse and menu-driven operations on iconic objects.

A key feature of the environment is an object-oriented approach which is implemented by the Object Management System (OMS) at the core of IMSE. Objects are derived from a common entity relationship model of the whole system and may, for example, be models, results, reports or collections of input values. The OMS provides functions that allow the tools, and therefore the user, to create and delete

[1]The IMSE project is ESPRIT2 project 2143, funded by the CEC. IMSE reports can be requested from BNR Europe Ltd, London Road, Harlow, Essex CM17 9NA, UK.

certain objects, group objects into directories, and to establish relational links between them.

IMSE currently supports three alternative performance modelling paradigms and has been designed to be open to new tools and techniques in the future. The paradigms currently included are timed, stochastic Petri networks, queueing networks and a tool based on the process oriented view of simulation. Each of these is based upon a previously existing tool: GreatSPN [20], QNAP2 [12] and PIT [21] respectively. Models are constructed by graphical editing tools, all based on a common data manipulation and graphical support system (SDMF) [22]. Models are therefore built in a diagrammatic format and automatically converted into an equivalent textual representation, suitable for use by a model solution engines. In general, the paradigm dependent details of using the model solution tools are hidden from the user.

Also included within the environment are tools for the static modelling of systems and the analysis of workloads. Again existing tools, *Sp* [9] and WAT [8], were used. The static models produced by *Sp* provide a hierarchical description of the system in terms of subsystems and offered and used services. The inclusion of WAT means that workload analyses analyses may be performed on external data to provide input for dynamic or static models. Alternatively, for validation purposes, such analyses may be performed on the output of models.

The final aspect of the IMSE environment is the tools provided to enable the user to make full use of the models created. These tools have been developed especially for the project. The Experimenter, handling the models in a paradigm independent way, allows the user to concentrate on the desired conditions for model execution and the treatment of results. There is also an Animator tool, driven by simulation traces which allows the user to visualise the execution, and a Reporter tool for the creation and collation of results and reports.

The IMSE graphical editor allows graphs to be composed hierarchically. That is, an entire graph edited may be represented as a primitive node within another graph, provided an appropriate interface is maintained. This hierarchical composition brings the powerful notions of data abstraction and modular construction into model construction. The internal structure of the subgraph can be changed without affecting the main graph, provided the interface remains consistent. The use of the OMS and common facilities for structuring and manipulating data mean that the submodel represented by the subgraph need not be developed in the same paradigm as the enclosing model. Coupled with the Experimenter this offers new possibilities for hybrid modelling.

4.2   HIT

The HIT system developed at the University of Dortmund has proved to be a very useful tool both academically and commercially. It is much more modest in its aims than the IMSE project and is aimed primarily at system designers.

It is a system description based modelling tool and the main motivation behind it was seen as the complexity of the systems being modelled. The diversification of hardware and software was increasing the importance of performance modelling whilst also making it more difficult. Also within this motivation was a recognition of the importance of being able to assess a system qualitatively as well as quantitatively, a key rôle of modelling being to enhance the understanding of the underlying system.

HIT was developed to address these issues and hoped to provide an interface to the performance modelling techniques which would allow modelling and evaluation to be done by those interested in the results of performance evaluation. It was aimed that these results should be accessible to the user without him being an expert in simulation, statistics, queueing theory, numerical analysis or related techniques. If this could be achieved the model would be developed by the person who understood the system and the objective of the study, thus eliminating the communication overhead incurred if the probelm is passed to a modelling specialist.

HIT incorporates a single means of specifying models regardless of how they are to be solved, and gives the user access to a variety of solution methods. However, HIT is not rigidly based upon one paradigm for model expression such as queueing networks, although when appropriate HIT models can be recognised as being of this form and advantage is taken of the available analytical solution techniques. This gives HIT greater freedom of expression and reflects the emphasis of the tool which is in consideration of systems and not models.

The approach taken by HIT is to provide the means to develop a structured specification of the system. Based upon common software engineering practices a hierarchical and modular view of the system is supported. The user-interface although still primarily textual, does provide a graphical editor, allowing the user to enter the structural information by means of diagrams.

An object management facility, OMA is incorporated into HIT and this supports the configuration of HIT models and experiments. Models may be parameterised so that the model designer builds a class of models which are later instantiated for particular use. This encourages the reuse of models.

HIT incorporates a variety of evaluation techniques:

- exact product-form solutions for "separable networks"

- approximate techniques for "large" separable networks

- numerical evaluation of Markov-cahin representations of general models

- stochastic discrete-event simulation with appropriate statical result estimation;

and allows the user to select the most appropriate one. Model specification is independent of the evaluation technique employed although there are different code generators depending on the model evelution technique selected, and the model is linked to the selected evaluation module. Clearly not all models specified are tractable by all resolution techniques.

The idea of experiments is supported by a control file describing what a particular experiment is concerned with.

HIT can be classed as a heterogeneous modelling tool, offering the user the opportunity of combined utilization of several analysis techniques in order to provide a comprehensive modelling tool. Unlike other heterogeneous modelling tools like QNAP2, HIT allows the user to directly incorporate one model within another, offering particularly strong support for DAT.

5 Issues for future work

The challenges for performance analysis are enormous, as systems of vastly increased complexity are the objects of its interest. Paradoxically the sheer increase in power and speed of computer and communications systems is increasing the need to be able to deliver that power to users. This means that performance analysis has more potential than ever to contribute to computer hardware and software engineering, but that traditional approaches are proving inadequate to the task.

One clear need is for faster and more accurate ways of solving models. In particular analytic and numerical solution techniques must become efficient and capable of handling larger classes of more complex models. This will give us the power, but only as an engine. The most promising development here is more general techniques for pre-analysis of submodels for substitution as aggregates in parent models. These extend existing approaches such as flow-equivalent servers and offer substantial performance gains and a means of combining different solution techniques in hierarchical models [23].

Since most model solution techniques depend on an underlying Markovian model, we should be free to choose ways of expressing such models at the user level, in order to suit the user. At present queueing networks are still dominant, although it is worth remembering that even they are only really some fifteen years or so old in this context. Petri nets are currently fashionable with researchers, but the real challenge is to start from the users' existing formalisms and map onto the solvers. HIT is an example of a move in this direction.

The existence of timed versions of process algebras, such as Timed CCS, widens the potential for expressing models in new ways and at the same time merging with qualitative techniques. Only preliminary work [24], [25] in this direction is known to the authors at present. The question is whether to find ways of directly solving timed process algebra models for performance results or ways of mapping existing performance formalisms onto timed algebras for functional results.

Finally, there is the challenge of how to support these techniques, along with existing ones. The efficient answer seems to be to create a genuine support environment, using lessons learned from current examples. The need for, and the benefits to be gained from, an object base and user interface tools seem to be proven by experiences with the IMSE. The potential for generic tools for experimentation, analysis and report generation are also hard to discount. The problems remaining are how to make it easy to share engines efficiently, which seems to require tools to be built explicitly to live in a shared environment, while allowing customisation and extensibility. The need to provide a better semantic framework for modelling seems to underlie all of this.

What is clear is that the stage of development of tools is only just entering the integration stage. The potential is undeniable and some of the achievements are significant, but the realisation of the full potential requires answers to all of the problems outlined here.

References

[1] Ören T.I. and J.C.M. Tam "An Expert Modelling and Simulation System on

Sun Workstations" in Proceedings of 2nd European Simulation Multiconference, Nice, June 1988.

[2] H. Beilner, J. Mäter, and N. Weißenberg., "Towards a Performance Modelling Environment: News on HIT", in *Proceedings of the Fourth International Conference on Modelling Techniques and Tools for Computer Evaluation*, Plenum Publishing, 1988.

[3] L. Golubchik, G. Rozenblat, W. Cheng, R. Muntz "The Tangram Modelling Environment", Proceedings of 5th International Conference on Modelling Techniques and Tools for Computer Performance Evaluation, Torino, February 1991, pp. 421-435, North Holland.

[4] R. Pooley 'The Integrated Modelling Support Environment", Proceedings of 5th International Conference on Modelling Techniques and Tools for Computer Performance Evaluation, Torino, February 1991, pp. 1-15, North Holland.

[5] C.U. Smith "Performance Engineering of Software Systems", Addison Wesley, New York, 1990.

[6] D. Ferrari, G. Serazzi. A. Zeigner "Measurement and Tuning of Computer Systems", Prentice Hall, Eaglewood Cliffs, 1983.

[7] M. Calzarossa and G. Serazzi "A software tool for the workload analysis" in N. Abu El Ata ed. Modelling Techniques and Tools for Performance Analysis (North-Holland, Amsterdam, 1985) 165-180.

[8] M.C. Calzarossa and L. Massala "WAT User Guide", IMSE Document R 4.2-4, 1991.

[9] Hughes P.H. *"Sp Principles"*, *STC Technical Report*, July 1988, STC Technology Ltd, Copthall House, Newcastle-under-Lyme, Staffordshire, UK (now BNR Europe Ltd, London Road, Harlow, Essex CM17 9NA, UK.)

[10] W. Dulz and S. Hofman, "Grammar Based Workload Modelling of Communication Systems", Proceedings of 5th International Conference on Modelling Techniques and Tools for Computer Performance Evaluation, Torino, February 1991, pp. 16-30, North Holland.

[11] B. Melamed and R.J.T. Morris "Visual Simulation: the Performance Analysis Workstation", IEEE Computer, Vol 18 No 8, pp. 87-94.

[12] M. Veran and D. Potier "QNAP 2: A Portable Environment for Queueing Systems Modelling", in *Modelling Techniques and Tools for Performance Analysis*, D. Potier(Ed), North Holland, 1985.

[13] R.P. Taylor and R.D. Hurrion, "Support Environments for discrete Event Simulation Experimentation", Proceedings of European Simulation Multiconference, Nice, June 1988, pp 242-8, SCS Europe.

[14] T.I. Ören., "Gest: A Modelling and Simulation Language Based on System Theoretic Concepts", in *Simulation and Model-Based Methodologies: An Integrative View*", T.I. Ören, B.P. Zeigler, and M.S. Elzas (Eds), Springer-Verlag 1984, pp 281-335.

[15] Zeigler B.P. "Theory of Modelling and Simulation", Krieger 1976.

[16] J. Hillston, R.J. Pooley and N. Stevenson "An Experimentation Facility Within the Integrated Modelling Support Environment" in Proc. of the UKSC Conference on Computer Simulation, Brighton, September 1990.

[17] J. Hillston "System Description Formalisms and Performance Evaluation", IMSE Document D4.4-3, Edinburgh University. 1991. In preparation.

[18] Pooley R.J. and M.W. Brown June 1988, "Automated modelling with the General Attributed (Directed) Graph Editing Tool", Proceeding of the European Simulation Multiconference, Nice, Society for Computer Simulation.

[19] G. Titterington and A. Thomas "The IMSE Object Management System", IMSE Document D2.1-5, 1990.

[20] G. Chiola "A Graphical Petri Net Tool for Performance Analysis" in Proceedings 3rd International Workshop on Modelling Techniques and Performance Evaluation, AFCET, Paris, March 1987.

[21] E.O. Barber "The Process Interaction Tool User Guide" IMSE Report R5.1-4, 1991.

[22] C. Uppal "The Design of the IMSE Standard Data Manipulation Facility", IMSE Document R2.2-3. 1990.

[23] "Numerical Solution Methods based on Structured Descriptions of Markovian Models", Proceedings of 5th International Conference on Modelling Techniques and Tools for Computer Performance Evaluation, Torino, February 1991, pp. 242-258, North Holland.

[24] U. Herzog "Formal Description, Time and Performance Analysis a Framework", University of Erlangen-Nürnberg Internal Report, September 1990.

[25] R. Pooley "Process Interaction Simulation Models as Timed CCS Models", in preparation.

A Statistical Approach to Finding Performance Models of Parallel Programs

Rosemary Candlin
Department of Computer Science

Peter Fisk
Department of Statistics

Neil Skilling
Department of Chemical Engineering
University of Edinburgh

Abstract

We wish to be able to predict the execution time of a parallel program from a knowledge of the values of certain parameters of the program and the environment in which it runs, and in order to do this, we have to discover which factors have an important influence on performance. In this paper, we describe a methodology for constructing synthetic programs corresponding to a particular model of parallel computation, simulating their execution, and analysing performance data using standard statistical techniques to estimate effects due to the various factors. The method is applied to a model representing parallel programs consisting of uniform concurrent processes, to show the kind of quantitative information that can be derived from experiments of this type

1 Introduction

It is well known that it is very difficult to predict the performance of a parallel program on a distributed-memory machine. The execution time depends on so many factors: the structure and dynamic properties of the program, the hardware characteristics of the machine, the topology of the hardware links and the decomposition and mapping of the program to the underlying machine. In this paper, we describe a methodology for obtaining a statistical model of the effects on performance of some of the important factors listed above.

We must distinguish between the general methodology that we propose, and the particular example that we present here. The former is quite general, and can be applied to any type of program and machine. In our actual example, we have made simplifying assumptions that may have to be altered in the light of further experiments.

We see this work as only a first step towards finding useful performance models of various classes of parallel programs. Our aim is to characterise a program in terms of a small number of parameters which can be derived from the program text, or measured by run-time monitoring, and to predict its execution time in a particular environment. The first stage of our work, therefore, is to see whether this is a valid approach, by exploring the way in which execution time varies as the parameter values vary, and seeing what conclusions we can draw.

Potentially this is a research area of great practical importance. Implementers of systems software for parallel machines need a firm basis on which to make decisions about compile- and run-time process allocation, and therefore require quantitative statistical information about the influence of various measurable factors on execution time. If appropriate strategies, which work well for programs in general, can be incorporated into compilers and operating systems, the individual programmer will be freed from the burden of having to worry about adapting applications programs to a particular machine. It seems essential to take this step, if parallel machines are to achieve a wide user base.

2   Choice of Approach

The execution time of a program depends in a very complicated way on the interaction between the program and the machine it runs on. It is not feasible to produce a realistic analytic or stochastic model of an individual program, and estimate the run time from that (at least not in our present state of knowledge of the dynamic behaviour of parallel programs). We have therefore taken an experimental approach. However, there are a number of disadvantages to running real programs on real machines. In the first place, it is very time consuming, and in the second, one is limited to the particular machines and programs that happen to be available. Since our aim is to be able to predict the run time of any arbitrary program, we have chosen to simulate the behaviour of a wide range of synthetic programs, so as to be able to sample from the whole of "program space". In outline, the work has involved the construction of the following software components:

- a performance-modelling system to simulate the execution of a program

- a synthetic program generator

- an experiment generator to allow model parameters to be systematically varied

A description of this experimental environment is provided in another paper submitted to this workshop [5].

In an experiment, we take a particular model of a class of parallel program, characterised by some given set of parameters. We then see how the execution time of the program depends on the values of the parameters by measuring the simulated execution time of a sample of synthetic programs. The results are then submitted to a standard statistical analysis.

The three system components mentioned above are now all in place, and we have started exploring statistical models. This work is still at an early stage but is already suggesting a number of interesting research directions.

3   Experimental Background

3.1   The MIMD Modelling System

As described in [5], MIMD allows a user to build a specific model of a particular machine, a particular program, and a particular placement. It has been used as an extended profiling tool, to evaluate the performance of user programs under different placements [10]. It can also be used to build synthetic programs which have given statistical properties. To take a simple case, we can model a program consisting of a set of parallel processes, each of which executes an endless loop within which the compute time is described by a normal distribution. We can now vary the parameters of the distribution and see how the behaviour of the program varies. We might start by seeing what happens in a program with identical processes, and then, in the light of results obtained, progress to programs where the processes have different mean compute times. As we explore, we find that some factors are unimportant, whereas others need to be studied at a wider set of values. The advantage of using synthetic programs is that we can provide test material easily and systematically.

3.2   Cost function

For a given program and computer, there is only one measure of interest: how long does the program take to complete? As will be seen later, our synthetic programs are non-terminating and we need another measure. We have chosen a measure related to the rate of computation: the average, over all the processes in the program and for a fixed simulation time, of the total compute time per process. By taking an average over a period of time which is long compared to an instruction execution time, we greatly reduce fluctuations due to the detailed order in which particular processes engage in particular actions and justify our assumption that useful conclusions can be derived from a model in which we ignore the effects of such ordering.

Our choice of cost function is provisional. Our assumption is that a good program mapping enables all processes to spend a high proportion of their time computing, but this may be an over-simplification for some programs with bottleneck processes. We need much more experimental data on real programs before we can be confident that, in general, the process average gives a good indication of overall performance.

3.3   Synthetic Program Generation

We have first to consider what aspects of the execution of a parallel program we wish to study, and justify our choice of model. We have to accept some limitations on the particular type of programs we can handle, just because there are a bewildering number of factors that might be of importance, and we cannot allow them all to vary at the same time. In arriving at our first proposed model we have restricted ourselves to programs whose parameters remain constant throughout their execution.

3.3.1   Type of program

Our main interest at the moment is in single-user programs running on distributed memory machines. Such programs can be decomposed into a number of concurrent processes exchanging information by message-passing. In our first model, we have confined ourselves to one level of parallelism, and represented the structure of the program in terms of a graph, whose nodes represent sequential processes and whose edges represent interprocess communication channels. This type of graph is a natural model for programs with a static structure, in particular for programs written in occam.

3.3.2 Model Parameters

We are looking for a description with the following properties:

- The parameter set should be small. This is not entirely our own choice, since it may turn out that the behaviour of the system is too complicated to describe in a simple way. Nonetheless, we should see how far we can go with a simple model.

- We should include only those factors which have a substantial effect on execution rate. We are not interested in factors that only make a difference of a few percent.

- The parameters should be easily measurable experimentally. It should be possible to derive them at compile-time, or measure them directly by monitoring the running program.

  Parameters have to represent two aspects of the program: its inherent structure (the size and shape of the graph) and its run-time behaviour (its computational and communications behaviour).

The synthetic program generator requires three components:

- a graph generator

- a method of mapping sets of weights onto each node and edge

- process templates (MIMD Simula procedures).

For a given combination of parameters, a program instance is produced in the following way:

1. a graph instance is drawn from the set of graphs of given size and connectivity

2. global (i.e. averaged over nodes and edges) weights are drawn from standard statistical distributions

3. node and edge weights, representing time-averaged properties for individual processes and channels, are derived from the global weights

4. the graph is "animated" according to specified process templates to simulate the execution of the program.

It can be seen that there are a number of implicit variables arising from the choice of distributions and process templates. Given our ignorance of the statistical properties of concurrent programs, the choice of the most appropriate distribution is difficult to make. We intend therefore to repeat our experiments using different distributions to see how much this choice influences our results.

Similarly, a process template describes a rather specific type of computation, and we shall need to run experiments with a number of different models.

4 First Experiments

4.1 Program Model

There has been a certain amount of experimental work on the performance of parallel programs, but no very precise conclusions have emerged. However, certain qualitative results seem to be established, one of which is that there is a very marked interaction between the topology of the program and that of the machine on which it runs. This is because communication costs can predominate if there is a lot of message traffic over long distances. For our first experiments, we have kept both machine and topology fixed, and have just explored the effects of changing inherent program parameters. We have in fact modelled a transputer- based array connected as a grid, since this will provide us with a machine model which is sufficiently typical to give interesting results. This model has been previously validated experimentally and gives timings which are accurate to within 5%.

Our first program model has the following characteristics:

- Its structure is represented by a regular graph, parameterised by N, the number of nodes, and c the connectivity.

- Each process has the same dynamic behaviour, and is represented by a process template like that below:

```
while true do begin
    delay for a random computation time
    send and receive messages on all connected channels
end
```

  The messages have a variable length which is drawn at random from an appropriate distribution. The communications protocol implements the occam blocking send, and results in a loose synchronization between processes.

- The amounts of computation and communication in the program are represented by (μ_c, σ_c) and (μ_m, σ_m), the mean and standard deviation averaged over the nodes and edges respectively of the computation time and the message length.

Each program is thus represented by six parameters $(N, c, \mu_c, \sigma_c, \mu_m, \sigma_m)$, which at first looks like an extreme simplification. Our purpose at this stage, however, is to establish the methodology, and see how far we can get with it. It is worth pointing out that this apparently limited model does represent an important class of program. Many programs exhibiting geometric parallelism are structured on distributed memory machines as an ensemble of similar processes, with each process carrying out one iteration of its loop, and then exchanging data with its neighbours before proceeding to the next iteration.

4.2 Sampling from program space

A particular combination of parameters represents a whole set of program instances, which may differ one from another in the following ways:

- their connection pattern

- the distribution of time-averaged process computation times over the nodes of the graph

- the distribution of time-averaged message lengths over the edges of the graph

- the interleaving in time of the activities of the various processes.

One of the purposes of our experiments will be to see how much variation can be attributed to differences between program instances with the same parameter combination, and how much arises from differences in parameter values. This will give an indication of the success of the model from the point of view of predicting the performance of an arbitrary program derived from the model. It will not, however, tell us if our model is realistic, from the point of view of predicting the performance of a real program: this will require experimental measurements of real programs on real machines.

4.3   Simulation Experiments

4.3.1   Input to an Experiment

Our experiments are generated by the experiment generator as mentioned previously. We have conducted a two-level, full factorial experiment varying the six parameters mentioned in Section 4.1.

We have used only one type of hardware for the initial experiments. It is a 4x4 connected mesh. We have also used just one process placement strategy.

4.3.2   Experimental Method

The experimental approach adopted is to conduct the same number k of simulation trials at each possible combination of the two extreme values chosen for each of the six parameters. For $k = 1$, this would require $2^6 = 64$ trials. This is a full factorial design replicated k times [11]. The result of each trial will be a measure of performance (average amount of computation per process). The value of this measure will differ from one trial to another at the same combination of the six parameter values because of the random selection of computation times and message lengths for the different nodes and edges. Thus, the outcome measurement will not be a constant function of the six parameter values.

The effect of each parameter on the outcome measurement can be determined from a suitable model of the relationship. The approach can be illustrated with two parameters. Let the two values for each parameter be represented as θ_{iL} and θ_{iU} for $i = 1, 2$. Here L denotes the lower and U the upper value. The mid-point of this range is $\bar{\theta}_i = (\theta_{iL} + \theta_{iU})/2$. As each parameter has been included in the experiment at two values only, the only relationship that can be described between outcome and one parameter is a straight line. This straight line may have a different slope and intersect for each value of the other parameter, but the relationship between intercept and slope can still only be represented by a straight line. We are thus led to the simplified form of relationship suggested by the structure of the design of the experiment:

$$y = \beta_0 + \beta_1\theta_1 + \beta_2\theta_2 + \beta_{12}\theta_1\theta_2 + \epsilon$$

with $\theta_i = \theta_{iL}$ or θ_{iU}. This may be rewritten in a variety of ways. One would be to replace θ_i by $\theta_i - \bar{\theta}_i$ with a corresponding change in the values of the β's. This has the advantage that the model may be represented as:

$$y = \gamma_0 + \gamma_1 z_1 + \gamma_2 z_2 + \gamma_{12} z_1 z_2 + \epsilon$$

with $z_i = \pm 1$, $\gamma_i = \beta_i^*(\theta_{iU} - \theta_{iL})/2$ for $i = 1, 2$ and $\gamma_{12} = \beta_{12}(\theta_{1U} - \theta_{1L})(\theta_{2U} - \theta_{2L})/4$. Here the β^*'s are linear functions of the β's. This is the statistical analysis of variance model which allows one to obtain estimates of the γ's by an orthogonal transformation of the means of the y's at each of the four combinations of parameter values. The same transformation allows the total sum of squared deviations of y's about their overall mean to be decomposed into separate quadratic forms for each of the γ estimates. The terms in the decomposition indicate the relative contribution of each term in the model to the overall variation in the outcome measurement.

The modelling described here can readily be extended to any number of parameters.

The first experiment to test the simulation procedure made very simple assumptions about the properties of programs. These assumptions were chosen for their convenience in operation rather than as realistic descriptions of a class of programs on a concurrent surface. The number of nodes N and the connectivity c were chosen to have one of two values. The amount of computation time at each node was assumed to follow a Normal distribution truncated at zero by rejecting any negative values produced by the Normal variate generator. The standard deviation at each node was assumed to take one of two values, but the mean was assumed to differ between nodes. Values of the mean were chosen at random from a Normal distribution with a fixed variance but one of two mean values. In a similar way, the message lengths on each edge were assumed to follow a Normal distribution with one of two values for the standard deviation but a different mean on each edge. The mean for each edge was selected from a Normal distribution with one of two values for the mean but a fixed variance. The values for the parameters were thus:

- Graph shape parameters

  N : 16 or 150 ; c : 4 or 15

- Computation time distribution

  σ_c : 5 or 50

  μ_c : selected from Normal distribution with standard deviation 5 and mean 500 or 500000

- Message length distribution

  σ_m : 0.01 or 0.1

  μ_m : selected from Normal distribution with standard deviation 0.01 and mean 1.0 or 10000

The experiment was conducted with two replicates ($k = 2$), and the results of the transformation of the 64 means giving "estimates" of the γ's are given in Table 5. The corresponding analysis of variance table is given in Table 2. The latter shows that the differences between the two replicates are very small relative to the estimate of the variability between program instances for the same parameter values (the error mean square). By contrast the differences between parameter settings (the "treatment" mean square) are very marked. Each estimate in Table 5, apart from that for the constant term γ_0, has the same standard error which is given at the foot of the table. The t-values given in this table are the ratio of estimates to standard error. Values of t less than 3 can be taken as a guide to those terms in the model which are of little importance.

The t-values for the terms involving the standard deviations for both the computation time and message length distributions are all markedly small relative to the rest. Even though some are greater than 3, the disparity between the t-values including and excluding these standard deviations suggests that they are of less practical importance than the other four parameters. It must be admitted that this could reflect the choice of values for the standard deviations in the trials rather than the relative influence of the standard deviations. This needs to be explored further. Table 3 gives analysis of variance table with the contributions in the treatment mean square of Table 2 from the two standard deviation parameters merged with the error sum of squares of that table. The mean square for the error in Table 3 is noticeably larger than that in Table 3 which may suggest caution in the abandonment of the standard errors as explanatory factors. Table 4 gives the γ estimates involving the remaining four parameters together with t-statistics using the standard error derived from the error variance in Table 3. It will be seen that the estimated values are unchanged from Table 5 (a feature of the experimental design and the model chosen). Ignoring those γ parameters which appear unimportant we end with a possible model for the computation done per process T given by:

$$
\begin{aligned}
T = {} & 2756415 - 115684z_2 + 1608756z_4 + 102908z_2z_4 + 39513z_2z_5 \\
& + 101265z_4z_5 - 26746z_2z_4z_5 - 263454z_6 - 3517z_2z_6 + 205955z_4z_6 \\
& + 6382z_2z_4z_6 - 29643z_2z_5z_6 + 58531z_4z_5z_6 + 26770z_2z_4z_5z_6 + e
\end{aligned}
$$

The plausibility of this model would need to be tested by further trials at other values of the six parameters.

The advantage of this approach is that it allows flexibility to explore the relative merits of alternative models as predictors of the performance of a program with particular characteristics on a specified type of machine. A great deal of information can be obtained with a limited input of resources in computational time. The approach can be extended to non-linear models by using factorials at three levels of some or all parameters. It can also be reduced by using fractional factorials when it can be appreciated that high-order terms, as mentioned above, can be assumed to be of no importance.

5 Future Work

There are a great many other aspects of parallel computations that can affect performance. We wish to apply the same techniques to studying the effect on performance of different machine models and different mapping heuristics.

There are also a number of different models of parallel programs that we intend to study, in particular programs which are irregular in their structure and run-time behaviour.

References

[1] Joe Phillips and Neil Skilling, *A Modelling Environment for Studying the Performance of Parallel Programs*, submitted to the Seventh Uk Computer and Telecommunications Performance Engineering Workshop, Edinburgh, $22^{nd}-24^{th}$ July 1991.

[2] Inmos Ltd, *Occam 2 Reference Manual*, Prentice Hall International, 1988.

[3] Inmos Ltd, *The Transputer Databook*, Inmos Ltd, Bristol, 1989.

[4] R. Candlin and N. Skilling, "A Modelling System for the Investigation of Parallel Program Performance", in *Proceedings of The Fifth International Conference on Modelling Techniques and Tools for Computer Performance Evaluation*, Turin, February 1991.

[5] N. Skilling, "**eg** - An Experiment Generator for Studying the Performance of Parallel Programs", Unpublished as yet, available from author.

[6] N. Skilling, *MIMD*, University of Edinburgh, Department of Chemical Engineering, Technical Report 1991-7.

[7] R. Candlin, T. Guilfoy and N. Skilling, "A Modelling System for Process-based Programs" in *Proceedings of the European Simulation Congress*, Edinburgh 1989.

[8] G.M. Birtwistle, *Discrete Event Modelling on Simula*, Macmillan, London, 1986.

[9] J. Phillips, "A Modelling Environment for Time-Varying Parallel Programs", Unpublished as yet, available from author.

[10] Rosemary Candlin, Qiangyi Luo and Neil Skilling, *The Investigation of Communication Patterns in Occam Programs*, Occam User Group Technical Meeting 11, 1989.

[11] W.G. Cochran and G.M. Cox, *Experimental Design*, J. Wiley, Second edition, 1957.

Table 1: Computation done per process, double replicate. Estimates of γ's for all factors and interactions in the model.

effects	estimates	t	effects	estimates	t
gm	2756415.1028	5577.90	$\sigma_m c$	-64.4783	-0.26
σ_m	1020.9299	4.13	$\mu_m c$	-3517.4314	-14.24
μ_m	-115684.6443	-468.20	$\sigma_m \mu_m c$	-496.4675	-2.01
$\sigma_m \mu_m$	-476.8015	-1.93	$\sigma_c c$	157.7888	0.64
σ_c	761.5525	3.08	$\sigma_m \sigma_c c$	392.1316	1.59
$\sigma_m \sigma_c$	546.4226	2.21	$\mu_m \sigma_c c$	-803.1789	-3.25
$\mu_m \sigma_c$	-238.3684	-0.96	$\sigma_m \mu_m \sigma_c c$	-977.2073	-3.95
$\sigma_m \mu_m \sigma_c$	-3.6897	-0.01	$\mu_c c$	205955.0867	833.54
μ_c	1608756.2527	6510.98	$\sigma_m \mu_c c$	69.3806	0.28
$\sigma_m \mu_c$	924.7381	3.74	$\mu_m \mu_c c$	6382.5760	25.83
$\mu_m \mu_c$	102908.1482	416.49	$\sigma_m \mu_m \mu_c c$	-477.0754	-1.93
$\sigma_m \mu_m \mu_c$	-484.8196	-1.96	$\sigma_c \mu_c c$	788.9339	3.19
$\sigma_c \mu_c$	189.5296	0.77	$\sigma_m \sigma_c \mu_c c$	590.6535	2.39
$\sigma_m \sigma_c \mu_c$	423.9719	1.72	$\mu_m \sigma_c \mu_c c$	-1157.6218	-4.69
$\mu_m \sigma_c \mu_c$	241.7082	0.98	$\sigma_m \mu_m \sigma_c \mu_c c$	-972.3747	-3.94
$\sigma_m \mu_m \sigma_c \mu_c$	18.4358	0.07	Nc	-476.4812	-1.93
N	-1130.9870	-4.58	σ_mNc	10.2561	0.04
σ_mN	-964.1209	-3.90	μ_mNc	-29643.4275	-119.97
μ_mN	39513.3401	159.92	$\sigma_m \mu_m$Nc	440.9660	1.78
$\sigma_m \mu_m$N	502.6673	2.03	σ_cNc	-112.4925	-0.46
σ_cN	-793.1702	-3.21	$\sigma_m \sigma_c$Nc	-375.7747	-1.52
$\sigma_m \sigma_c$N	-514.7141	-2.08	$\mu_m \sigma_c$Nc	818.5593	3.31
$\mu_m \sigma_c$N	197.4831	0.80	$\sigma_m \mu_m \sigma_c$Nc	821.3669	3.32
$\sigma_m \mu_m \sigma_c$N	17.0369	0.07	μ_cNc	58531.5476	236.89
μ_cN	101265.4753	409.84	$\sigma_m \mu_c$Nc	-15.1272	-0.06
$\sigma_m \mu_c$N	-981.6409	-3.97	$\mu_m \mu_c$Nc	26770.7830	108.35
$\mu_m \mu_c$N	-26746.3440	-108.25	$\sigma_m \mu_m \mu_c$Nc	532.5769	2.16
$\sigma_m \mu_m \mu_c$N	458.8913	1.86	$\sigma_c \mu_c$Nc	-882.0739	-3.57
$\sigma_c \mu_c$N	-197.8181	-0.80	$\sigma_m \sigma_c \mu_c$Nc	-606.9792	-2.46
$\sigma_m \sigma_c \mu_c$N	-455.7742	-1.84	$\mu_m \sigma_c \mu_c$Nc	1142.4914	4.62
$\mu_m \sigma_c \mu_c$N	-200.8230	-0.81	$\sigma_m \mu_m \sigma_c \mu c$Nc	1128.2151	4.57
$\sigma_m \mu_m \sigma_c \mu_c$N	-31.8454	-0.13			
c	-263454.4968	-1066.26	Standard Error = 247.085		

Table 2: Analysis of variance table for the model with all factors and interactions included

	df	ss	ms	f
Replicates	1	1.135e04	1.135e04	7.26e-4
Treatment	63	3.509e14	5.570e12	3.56e05
Error	63	9.846e08	1.563e07	
Total	127	3.509e14		

df = degrees of freedom
ss = sum of squares
ms = mean square
f = ratio ms and ms error

Table 3: Analysis of variance table with standard deviation parameters excluded from model

	df	ss	ms	f
Replicates	1	1.135e04	1.135e04	0.084
Treatment	15	3.509e14	2.339e13	7.79e5
Error	111	3.335e09	3.005e07	
Total	127	3.509e14		

df = degrees of freedom
ss = sum of squares
ms = mean square
f = ratio ms and ms error

Table 4: Estimates of γ's excluding two standard deviation parameters

effects	estimates	t
μ_m	-115684.6443	-238.7529018
μ_c	1608756.2525	3320.1919410
$\mu_m\mu_c$	102908.1482	212.3844453
N	-1130.9871	-2.3341598
μ_mN	39513.3401	81.5486331
μ_cN	101265.4753	208.9942554
$\mu_m\mu_c$N	-26746.3439	-55.1997828
c	-263454.4968	-543.7240699
μ_mc	-3517.4314	-7.2593641
μ_cc	205955.0867	425.0553295
$\mu_m\mu_c$c	6382.5759	13.1725220
Nc	-476.4812	-0.9833740
μ_mNc	-29643.4275	-61.1788572
μ_cNc	58531.5476	120.7988919
$\mu_m\mu_c$Nc	26770.7829	55.2502206

Standard error= 484.537

A Modelling Environment for Studying the Performance of Parallel Programs

Joe Phillips
Department of Computer Science

Neil Skilling
Department of Chemical Engineering

King's Buildings
Edinburgh University
Mayfield Road
Edinburgh
EH9 3JL
United Kingdom

Abstract

As a first step towards providing automatic compile- and run-time process allocation decisions for distributed memory multiprocessors we are investigating appropriate performance models for various classes of message passing parallel programs. This paper describes a modelling environment within which experiments concerning the behaviour of message passing parallel programs can be carried out.

1 Introduction

The lack of adequate programming and run-time environments for distributed memory multiprocessors has tended to restrict the usage of such machines. For example, users generally have to explicitly map applications onto target architectures at compile-time, retaining the same mapping for the entirety of the run. It would be desirable for the systems software to be able to make sensible compile- and run-time process allocation decisions. This would both ease the burden on the programmer and allow the development of architecture independent applications. As a first step in this direction we are interested in finding appropriate performance models for various classes of message passing parallel programs.

We intend to characterise parallel programs in terms of a relatively small number of parameters. By undertaking a quantitative statistical analysis of *synthetic* parallel programs we hope to gain insight into the relationships between program parameters and program execution times. Ultimately we hope that our models will allow us to predict an arbitrary program's execution time in a particular environment.

We represent parallel programs as labelled, directed graphs, which are representative of a particular program shape. The nodes in the graph represent processes and the arcs represent channels. The labels define process and channel behaviours, arcs (and nodes) with the same labels represent channels (and processes) that exhibit the same behaviour. A label consists of a name together with a list of numeric values. No meanings are attached to labels until the graph is interpreted as a parallel program specification.

We have decided to use synthetic programs rather than real ones so that we can fully explore program space. Real programs are difficult to parameterise and vary systematically. At the moment we have two techniques available for generating *regular* random program graphs, [1] and [2]. The graphs are regular in the sense that all nodes are of the same degree. This is a restriction and we intend to extend our investigations to include irregular random graphs at a later date. The extent to which these graphs accurately represent the structure of real programs is an interesting research area in its own right. However, by designing our experiments carefully, we can ensure that real programs fall somewhere within the range of parameters explored.

It is with single user, message passing, process based computations on distributed memory multiprocessors that we are concerned. At the moment we are concentrating on simulations based on occam [4] and Transputers [5] although we hope that our results will be of more general significance. These resources are readily available at Edinburgh [1]. Occam has the advantage of being a relatively simple language and Transputers are widely used both in research and in industry.

This paper describes a modelling environment within which experiments concerning the behaviour of message passing parallel programs can be carried out. The way in which one might go about analysing the results of such experiments is described elsewhere [3].

2 The Experimental Framework

We define an experiment as the task of investigating the effects of varying certain parameters characterising a parallel program on its run time behaviour. An experimental framework has been constructed to aid the definition, generation and execution of such experiments. There are three components to this system,

- Experiment Definition

- Experiment Generator

- Modelling Engine

The relationship between the components is illustrated in Figure 1. Firstly we have to define what we want to model, this is the task of the experiment definition. The experiment definition states in a formal manner which parameters are to be varied and how they are to be varied in the experiment. A single experiment definition

---

[1]Namely the Edinburgh Concurrent Supercomputer (ECS). The ECS is a large Meiko Computing Surface housing some 400 Inmos T800 Transputers located in the Edinburgh Parallel Computing Centre (EPCC).

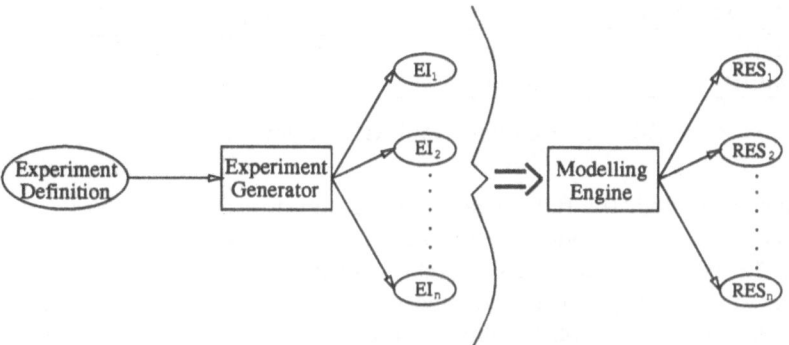

Figure 1: The Experimental Framework

would usually define a number of programs to be modelled. Secondly the experiment generator reads the experiment definition and produces a series of Execution Instances (EIs), one for each program to be modelled. Each EI contains all the information required to model a single parallel program. Finally, a modelling engine models the programs defined by the EIs in a particular *domain*. A corresponding series of result files are generated. These components are now examined in greater detail.

2.1 The Experiment Definition

An Experiment Definition Language (EDL) has been defined to allow experiments to be specified in a convenient manner.

The EDL allows one to specify at a reasonably high level all the components defining a parallel program, namely the program graph, the hardware graph and a mapping between the two. An EDL script can be split into two parts - there is the part that is concerned with specifying the structure of the EIs (i.e. *what* is to be modelled) and there is the part that is concerned with controlling the actions of the modelling engine.

2.1.1 Execution Parameters

The execution parameters define an experiment by specifying a set of parallel programs to be modelled. Within an EDL script certain execution parameters can be varied. By considering all possible combinations of variable parameters we can generate a set of parallel programs. A variable parameter can be a fixed value e.g. 7, an arbitrary list e.g. [1.5, 4.75, 5.25] or a list of values with a constant difference between them e.g. 3 to 5 step 1. This approach is particularly useful in factorial type experiments. An EDL script with no varying parameters does not strictly speaking define an experiment but might be useful for test purposes.

Suppose that we have a description of a parallel program and wish to investigate what would happen if we made some changes to it. We can describe these changes using an EDL script, see Figure 2 for an example.

Lines 3-8 define the graph parameters. First of all the structure of the program graph is specified. In our example the Redfield random regular graph generation

Generation Technique	Keyword	Degree Range	Node Range
Redfield[1]	Redfield	3-5	0-160
Random Regular[2]	RRgen	0-30	0-160

Table 1: Random Graph Generation Strategies

technique is selected, the full range of options are presented in Table 1. Then the degree and number of nodes are specified. The number of nodes will be varied between 12 and 24 in steps of 4 thereby allowing a set of program graphs to be investigated. The degree could also be varied. Alternatively one could supply a user defined program graph in which case the number of nodes and the degree of the program graph would be held constant for the duration of the experiment. A graphical front end exists to aid the construction of such graphs [6].

The hardware is defined to be a 4 by 3 mesh. A number of topologies are available i.e. mesh, hypercube, star, ring, pipeline, tree, random or user defined. The hardware cannot be varied within a single experiment at the moment but this could easily be implemented in the future.

Lines 10-15 give the process template definitions. A process template has a name and a set of named numeric parameters which characterise a particular type of process behaviour. The numeric parameters can be of type *integer* or *double* and can be varied. In our example the process template **proc1** has two parameters, one of which is varied over the experiment and one of which remains constant. No meanings are attributed to the parameters at this stage, this is the task of the modelling engine. Any number of process templates can be defined. Channel templates are defined in a similar manner (lines 17-21).

Lines 23-26 allocate process template definitions to nodes of the program graph. Process template definitions are assigned randomly in the proportions specified. In our example 70% of nodes receive the behaviour defined by **proc1** and 30% receive that defined by **proc2**. Lines 28-30 allocate channel template definitions to arcs of the program graph in a similar manner. Eventually it will be possible to specify exactly which nodes and arcs are to receive which definitions by specifying a list of node (or arc) labels instead of a percentage. This feature would be useful when used in conjunction with user defined program graphs.

Lines 32-34 define the placement algorithm to be used. Two are available at the moment, strategy 1 is a round robin placement and strategy 2 is a random placement. These algorithms are coded in the C programming language and so can easily be modified, or new ones added.

2.1.2 Modelling Parameters

The modelling parameters tend to be specific to a particular domain and non-varying. They are used to control the way in which EIs are executed by the modelling engine. The modelling parameters in our example (lines 36-41) are tailored towards a MIMD based modelling engine. MIMD is a package for simulating message passing parallel programs being executed on a distributed memory architecture. The parameters specified in our example instruct the modelling engine to run each simulation for 20 million clock cycles of simulated machine time with no tracing information.

```
1   Begin Experiment
2
3   Begin Graph Parameters
4     Graph Type redfield
5     Degree 4
6     Number Nodes 12 to 24 step 4
7     Hardware mesh 4 3
8   End Graph Parameters
9
10  Begin Define Processes
11    proc1 {
12    Int Param1 [500, 800, 1000]
13    Double Param2 12.5 }
14    proc2 { .....
15  End Define Processes
16
17  Begin Define Channels
18    chan1 {
19    Int Param1 1000 to 3000 step 500
20    Int Param2 50 }
21  End Define Channels
22
23  Begin Allocate Processes
24    map proc1 to 70.0 %
25    map proc2 to 30.0 %
26  End Allocate Processes
27
28  Begin Allocate Channels
29    map chan1 to 100.0 %
30  End Allocate Channels
31
32  Begin Placement
33    Algorithm 1
34  End Placement
35
36  Begin Modelling Parameters
37    Text Domain "MIMD"
38    Double Simulation_Time 20000000.0
39    Int Simulation_Runs 1
40    Boolean Tracing false
41  End Modelling Parameters
42
43  End Experiment
```

Figure 2: An Experiment Definition

2.2 The Experiment Generator

The Experiment Generator (**eg**) is a tool that allows the generation of *domain independent* experiments from a EDL script. It generates a series of EIs, one for each of the possible combinations of varying parameters defined in the EDL script. For example the script in Figure 2 would produce $4 \times 3 = 12$ EIs. The EIs are simply files of numbers and text describing the program to be modelled. Any modelling parameters specified are simply appended to each of the EIs, they do not alter the generation process in any way.

One should note that it is very easy to generate large numbers of EIs. This is because the experiment generator always produces a full factorial experiment based on all the parameters which are varying. At the moment there is no way to restrict the particular combinations of parameters which are considered.

The experiment generator runs on Sun workstations.

2.3 The Modelling Engine

The modelling engine takes an EI as its input data. It must in some way model the execution of the program described by the EI (i.e. the software, the hardware and and a mapping between the two) and report its results which can then be processed and combined as part of the whole experiment. The actions of the modelling engine can be controlled by the modelling parameters.

The modelling engine can be a real parallel program which is able to emulate the parallel computation specified by the EI in a real domain, for example the ECS. Alternatively the modelling engine can be a simulation system which takes the EI and simulates the parallel computation specified, for example MIMD.

As we have seen an EI is a domain independent representation of a parallel program i.e. a program graph, a hardware graph and a mapping between the two. It is only at the modelling stage that the process and channel template definitions are interpreted. For example, some of the values associated with a process could be interpreted as the computational intensity of that process. Some of the values associated with a channel might be interpreted as the mean length of messages passed down that channel.

A single EI can be modelled in a number of different domains as well as in the same domain by different modelling engines. Several modelling engines have been written using the simulation language MIMD. In the following sections we give an overview of MIMD and describe two modelling engines that have been constructed to deal with distinct classes of parallel programs.

3 The MIMD Simulation System

MIMD [7] is a Multiple Instruction stream, Multiple Data stream computer simulation system which simulates the execution of arbitrary message passing parallel programs on user defined distributed memory multiprocessors. A full description of the MIMD package and its validation for occam programs running on the Meiko Computing Surface can be found in [8].

The system has been designed to obtain a comprehensive set of statistics regarding the performance of *correct* parallel programs, as a result features such as deadlock detection are not provided. MIMD *is not* an instruction-level simulator, rather it is concerned with adequately representing and simulating the computational and communications characteristics of parallel programs. The programs simulated have no meaning in the usual sense of the word, they merely represent certain *patterns of activity*.

MIMD is built on top of DEMOS [9] and Simula and runs on Sun workstations. DEMOS is an extension to Simula containing class definitions for modelling discrete event simulations. MIMD provides new classes which are appropriate for modelling the execution of message passing parallel programs on distributed memory architectures. At the moment MIMD is tailored towards simulating occam programs on Transputer machines. It would be a relatively easy task to extend the range of languages and machines supported by defining new subclasses to handle different communications protocols and scheduling strategies. The standard facilities provided by MIMD can be augmented where necessary with the full power of the Simula language. A global snapshot of the simulated machine can be taken at any time without interfering with the simulation taking place, this would not be possible in a truly distributed implementation.

Within an MIMD program objects representing the components of a parallel computation must be declared i.e hardware, software and a mapping between the two.

The hardware is represented by an undirected graph of homogeneous processors joined by hard links. Processors and links are treated as resources which may be claimed and released as and when they are required. Objects representing the processors must be declared and wired into the desired configuration. This can be done by using one of the standard topologies provided within MIMD or by defining an arbitrary topology.

The software is represented by a directed graph of processes joined by unidirectional channels. These processes can only engage in four types of behaviour, compute, sleep, send a message and receive a message. A *compute(n)* statement places a process in the CPU scheduling queue with a requirement for n cycles of CPU time. A *sleep(n)* statement sends a process to sleep for n seconds. A *send(C_i, n)* statement sends a message of length n bytes down channel C_i. A virtual channel mechanism has been implemented so the message may have to pass through intermediate processors on its way to its destination. The occam soft channel protocol applies to these virtual channels i.e. synchronous, point to point communications. A *receive(C_i)* statement receives a message on channel C_i. These statements allow different patterns of activity in message passing parallel programs to be adequately modelled.

The actual time taken for a computation or message transfer is usually longer than the value specified due to competition for resources. The computational power of the processors and the bandwidth of paths between processors are resources which have to be shared between the competing processes of the concurrent computation. This is achieved by time-slicing processors and queuing messages at links. The mechanisms of process scheduling and message passing are modelled explicitly at a fairly low level. Time slicing occurs as it would in a real system by maintaining a queue of active processes for each processor. Inter-processor message hops are explicitly modelled by passing the data through communications processes running on each processor. A message has to compete for both CPU time on the processors it passes through and message transfer time on the hard links separating those processors.

Finally processes are mapped to processors by calling procedures that set up suitable objects describing the mapping. The execution of the resulting Simula program corresponds to the simulation of the parallel program that we wish to model.

The MIMD system collects a wealth of hardware and software statistics which may be examined in a post-mortem manner. They include information on processor usage, link usage, process activity and channel activity.

4 Examples of Parallel Program Models

This section explains how we can characterise and represent two distinct classes of message passing parallel programs as directed, labelled graphs.

4.1 Static Regular Parallel Programs

A simple type of parallel program is one where each process is identical. This means that each process follows the same instruction path. One of the simplest but realistic paths that a process can take is that of an endless loop made up of a computation phase and a communication phase. This is the type of process we call a static regular process. Each program graph is a fixed regular graph and each process follows a simple loop.

At each iteration it computes for a time period selected from a distribution function and then it communicates with each other process it is connected to. The size of the packet which is sent or received is also selected from a distribution. This class of parallel program is very common. It is known either as a scattered spatial decomposition program or as a geometric decomposition program. It is quite common to have a real program made up of a set of identical processes which swap boundary values.

Each process needs very little information. It needs to know to which other processes it is connected and how it should decide on the length of its computation phase and communication packets. These are passed as simple numbers to each process. It then uses these numbers as parameters to a statistical distribution to select the values at each iteration of the loop.

Say that we wish to choose the length of the computation phase from a normal distribution then the process template from the EDL script looks like this

```
Begin Define Processes process1
 {
 double mean_compute [ 1000, 2000 ]
 double sd_compute    10 }
```

This suggests that we wish to generate two experiments. One with the mean length of the computation twice the size of the other. In this way we can create simple experiments.

Similar templates are made for the channels.

4.2 Time-Varying Parallel Programs

Many parallel programs exhibit non-uniform, non-predictable computational and communications patterns to a significant degree. For example, parallel discrete event simulations, data dependent applications, particle dynamics and neural net simulations. We call these programs *time-varying* parallel programs.

We characterise time-varying parallel programs by using function specifications to describe a program's behaviour. These *behavioural functions* define a program's (i.e. its process and channel) actions over time so that non-uniform, time dependent behaviour can be specified. Each node in a program graph has four behavioural functions mapped to it:

- **Granularity Function**

  This function defines the atomic unit of work for a process. Consequently it specifies the frequency with which the process *considers* carrying out message transfers on its incoming and outgoing channels.

- **Compute-Sleep Ratio Function**

  This function defines how computationally intensive a process is. If the values returned are close to 1 then the process will tend to compute whenever it can. As the values returned decrease the probability of the process sleeping rather than computing for a particular time period increases.

- **Code Function**

  This function returns a value indicating the amount of code store occupied by a process. This value is likely to be constant for the duration of the program.

- **Workspace Function**

  This function returns the amount of memory that a process's data occupies. This value will be a constant in languages such as occam, although it could vary in languages that support dynamic storage allocation.

Each arc in a program graph has two behavioural functions mapped to it:

- **Gap Function**

  This function defines the minimum gap between message transfers on a channel. The smaller the values produced the busier the channel will be.

- **Length Function**

  This function defines the length of messages that are sent on a particular channel. High values indicate long messages.

All the functions described above take a single parameter, the *current* simulation time. They all return a single value which is interpreted according to the context of the function. By selecting appropriate behavioural functions and allocating them carefully one should be able to generate any patterns of activity desired. It is envisaged that (initially anyway) one would use relatively simple functions e.g. normal distributions, step functions and constants. In addition these functions would be distributed in a constrained manner e.g. all nodes would be allocated the same **compute-sleep** function. As one's understanding increased more sophisticated functions could be introduced (e.g. periodic and multi-phased functions) and these functions could be allocated in more complex ways. A controlled approach to function specification will allow parameter space to be fully explored.

5 A Modelling Engine For Static Regular Parallel Programs

This modelling engine is a very simple device. It is a short MIMD program which reads an EI and simulates that particular combination of labelled program graph and hardware combination. In order to do this it reads the EI and creates a new process with the correct parameter values found in the file. This is simple to do since Simula is an object oriented language and so new **processes** or **processors** can be created as desired inside the MIMD system.

Say that we are using the process which chooses its next computation phase from a normal distribution. Each process is passed the mean and standard deviation for the distribution. The following pseudo-code describes the execution path of each process.

```
subprocess process1(mean,sd)
double mean,sd;
begin
   while true do
   begin
     compute_period = normal(mean,sd);
     compute(compute_period);
     communicate on channels;
   end
end
```

6 A Modelling Engine For Time-Varying Parallel Programs

A modelling engine has been constructed to simulate time-varying programs in the MIMD domain. Before a simulation can begin the program specification has to be correctly interpreted. This process is described below.

6.1 Interpreting Time-Varying Parallel Program Specifications

When presented with a program graph labelled with process and channel template definitions we have to decide how to interpret those definitions in terms of behavioural functions. Figure 3 illustrates how this is done.

The EIs typically contain entries of the form, **chan3 500 10 20 5**. This is a channel template definition that has 4 parameters. In addition there are process template definitions. We have to supply a *function mapping script* that tells the system how to interpret process and channel templates in terms of behavioural

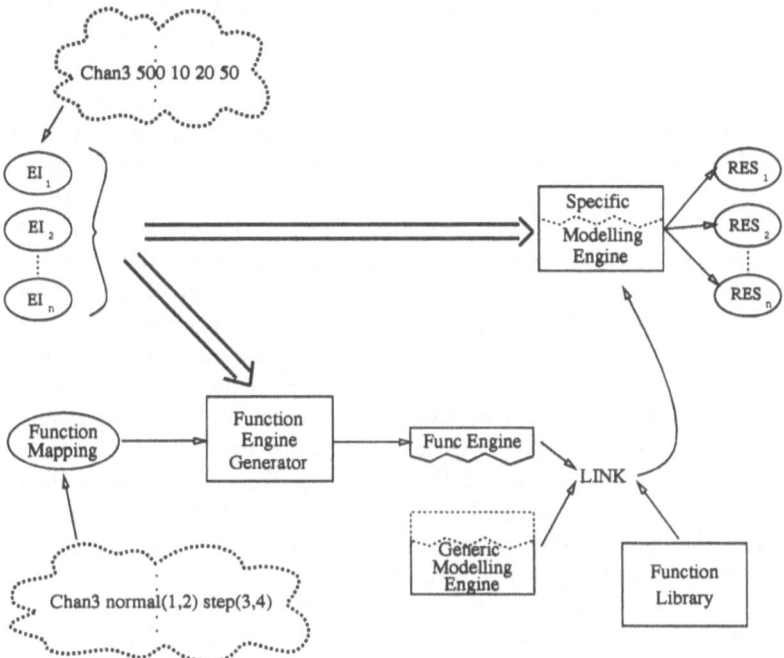

Figure 3: Interpreting Time-Varying Parallel Program Specifications

functions. For example the function mapping script might contain the line, **chan3 normal(1,2) step(3,4)**. This would explain to the system that when encountering labels with the name **chan3** that parameters 1 and 2 should be interpreted as values to be used to create an instance of the behavioural function called **normal**. Similarly parameters 3 and 4 should be interpreted as values to be used to create an instance of the behavioural function called **step**. The first function would be the inter-message **gap** function and the second would be the message **length** function. A similar procedure would be followed for process template definitions. All the behavioural functions referenced must be defined in a *Function Library*.

Figure 3 shows how the function mapping script and EIs are processed by the *Function Engine Generator* to produce a *Function Engine* containing all the behavioural functions required for a particular experiment. This function engine is then linked with the *generic* modelling engine and the function library to produce a modelling engine specific to this experiment. The EIs are then processed to produce the result files.

6.2   The Model Of Computation

This section describes a model of computation suitable for modelling time-varying occam type of programs. When generalising the behaviour of these types of program three realism criteria can be identified: there should be blocking reads, there should be blocking writes, programs should be deadlock free.

A program obeying the model of computation cycles through three states. Firstly a process *Computes* or *Sleeps* (depending on its **compute-sleep ratio** function) for a period determined by its **granularity** function. Secondly it polls its incoming and outgoing channels (in a random order) to see if any *Sends* or *Receives* have been initiated by a remote process. If so, a *Receive* or *Send* is invoked on the appropriate channel. The random order of channel processing ensures that no channels get preferential treatment. This prevents any unfair patterns of activity emerging.

In the last of the three states the process once again polls its incoming and outgoing channels in a random order to see if any message transfers are due. The **gap** function for a channel determines the frequency of messages. If a message is due it attempts to initiate a *Send* or *Receive*, depending on the direction of the channel in question. The lengths of messages sent on a channel are determined by the appropriate message **length** function. This attempt only succeeds if the remote process is *safe*, otherwise it is deferred. When a process is in one of the first two states it is considered to be *safe*. This mechanism prevents deadlock from occurring because no dependency cycles can evolve (i.e. where each process is blocked on its successor in the cycle). Why can no dependency cycles evolve? Consider the creation of such a cycle beginning at P_0, namely $P_0, P_1, ..., P_i$. To begin the cycle P_0 initiates a message transfer with P_1 and becomes blocked. Now, assuming that the remainder of the cycle arises, P_j can never close the cycle by initiating a message transfer with P_0 because P_0 is not in the *safe* state.

When a message transfer becomes due on a particular channel it is uncertain whether the sender or the receiver will initiate that transfer. This depends upon who reaches their initiation phase first and whether or not the other process is in a favourable state. If the sender initiates the transfer we get a blocking send and if the receiver initiates it we get a blocking receive. So we can see that all three realism criteria have been met.

6.3 Analysis

What class of parallel programs are simulated by the model of computation described above? Basically speaking the programs simulated correspond to those parallel programs in which message transfers on different channels are entirely independent. For example, in a real program it might be the case that a message transfer between P_i and P_j is always followed by a message transfer between P_j and P_k. This sort of behaviour cannot be specified in our model of computation. This is acceptable since for our purposes we are not concerned with the micro-ordering of channel transfers. The sequence of transfers $C_1, C_1, C_2, C_1, C_3, C_2, C_2$ is close enough to to $C_2, C_3, C_2, C_1, C_2, C_1, C_1$ from the statistical point of view, if one sums the transfers over a long enough time period. The two sequences are not entirely equivalent because of the differing synchronisation delays that would be incurred. The closeness (in the statistical sense) of these and similar sequences of channel usage could be tested by experiment.

7 Conclusions and Future Work

We have presented a modelling environment suitable for carrying out investigations into the general statistical properties of parallel programs. Experiments can be easily defined and the corresponding programs generated in a manner that is independent both of the characteristics of the programs being explored and the domain in which they are to be modelled. The user has total control of how any programs generated are to be modelled. Modelling engines for simulating static regular and time-varying parallel programs in the MIMD domain have been presented.

Future work includes investigation into placement strategies through the use of the experiment generator on various types of parallel programs. We also intend to carry out a comparison and analysis of process migration strategies for time-varying parallel programs in order to establish any relationships between program structure and migration strategy performance. Some of the questions that we would like to answer are, what characteristics of a program make it behave well under a particular process migration strategy? Also, what factors effect the performance of various process migration strategies when faced with a candidate program?

References

[1] E.M. Palmer, *Graphical Evolution : An Introduction to the Theory of Random Graphs*, New York : Wiley Interscience Series, 1985.

[2] M.R. Jerrum and A.J. Sinclair, "Fast uniform Generation of Regular Graphs", Internal report, Edinburgh University, Department of Computer Science, CSR-281-88.

[3] R. Candlin, P. Fisk and N. Skilling, *A Statistical Approach to Finding Performance Models of Parallel Programs,*presented at the Seventh UK Computer and Telecommunications Workshop, Edinburgh, July 1991.

[4] Inmos Ltd, *Occam 2 Reference Manual*, Prentice Hall International, 1988.

[5] Inmos Ltd, *The Transputer Databook*, Inmos Ltd, Bristol, 1989.

[6] R. Candlin and N. Skilling, "A Modelling System for the Investigation of Parallel Program Performance", in *Proceedings of The Fifth International Conference on Modelling Techniques and Tools for Computer Performance Evaluation*, Turin, February 1991.

[7] N. Skilling, *MIMD*, University of Edinburgh, Department of Chemical Engineering, Internal Report.

[8] R. Candlin, T. Guilfoy and N. Skilling, "A Modelling System for Process-based Programs" in *Proceedings of the European Simulation Congress*, Edinburgh 1989.

[9] G.M. Birtwistle, *Discrete Event Modelling on Simula*, Macmillan, London, 1986.

On the Performance Prediction of PRAM Simulating Models

G. Chochia *
Edinburgh Parallel Computing Centre
The University of Edinburgh
James Clerk Maxwell Building, Kings Buildings, Edinburgh, EH9 3JZ.

Abstract

This paper considers performance prediction for various computer networks in terms of the number of references made to the memory. Analytical solutions were obtained for the models with multiple requests to the memory and parallel access to one memory bank with a Negative Exponential distribution of requests generated by CPUs. Numerical simulation was carried out for Hypercube, Cube Connected Cycle and Torus processor networks. The distributed parallel architectures investigated could be viewed as PRAM simulating models. The analytical solutions are compared to numerical simulation results.

1 Introduction

The Parallel Random-Access Machine (PRAM) model assumes that n processors operating in parallel share a common memory. A large number of parallel algorithms have been formulated [1] in terms of these models. Unfortunately present technology does not allow us to implement PRAM models with a large n. The idea of simulating the PRAM on distributed memory parallel architectures was proposed [2] due to the desire to hide at the hardware level the difficulty of programming distributed memory parallel computers on one hand and to solve the problem of multiple accesses to shared memory on the other. There are several subclasses of these models that differ on how the memory access conflicts are resolved.

We use three basic elements to represent a computer network: these are CPUs, links and memories. Each CPU generates requests to the memory that are distributed according to the Negative Exponential (NE) law. We have chosen this type of distribution in the expectation that it would give a satisfactory approximation to the requests done to the memory for a large class of computers, and because it has important properties that are used in the theoretical analysis. The links and the memories are considered in the models as servers with S_{link} and S_{mem} service times. The case when only one request is allowed to be generated per CPU, which is idle until it comes back, we define as *the single-request model*. The more general case, when multiple requests are generated by one CPU, is classified as

---

*on sabbatical from the Institute for Space Research, Profsoyuznaja 84, Moscow, USSR

a multi-request model. Single-request models may serve to describe the sequential computers and networks of sequential computers. The multi-request models can be viewed as modelling vector machines and the networks built out of them.

To obtain a mathematical description of the network a Queueing Model [3] can be used. Queueing Theory deals with the equations describing the probability of finding the model in the particular state at a given time t. When $t \to \infty$ the probabilities are considered to be independent of time and the model is said have reached a steady state. The equations for the steady state probabilities are known as *balance equations*. In this paper we obtain the equations for the probability of entering a particular state during the time necessary for the request to be generated, reach the memory and come back, called *the reference cycle*. These equations can be used to find the number of references made to the memory in a given period of time.

In Sections 2 to 6 we find the analytical solutions for the multi-request single-CPU model and single-request parallel shared memory model, which serve as qualitative signposts for more complex parallel computer systems. To obtain quantitative results for complex networks we use numerical simulation. These simulations were implemented using Simula [4] with the DEMOS [5] (Discrete Event Modelling On Simula) package. The different architectures are built from identical basic blocks whose structure is described in Section 7. The probabilistic routing used during simulation is also introduced in that section. In Section 8 we consider average delays for the return of a request to memory, and in Section 9 we give a comparison of the networks simulated.

2   One CPU Model

We start our consideration with the simplest case: 1 CPU unit, 1 memory unit. The CPU generates requests to the memory which are distributed according NE law, with a probability density function (pdf):

$$p = \mu e^{-\mu t}, \tag{1}$$

where $\frac{1}{\mu}$ is the mean value. The CPU waits until the request has been satisfied before generating a new one. The period T of a reference cycle is given by

$$T = S + \xi,$$

where ξ is a random variable drawn from (1) and S is a memory access time or service time. Averaging this equation we obtain:

$$<T> = S + \frac{1}{\mu},$$

where $<T>$ is the average CPU period. Therefore, the average number of references to the memory observed in time T_{sim} is:

$$R = \frac{T_{sim}}{\left(S + \frac{1}{\mu}\right)}. \tag{2}$$

In fact these formulae hold for any distribution having mean value $\frac{1}{\mu}$. During this cyclic process the memory unit exists in one of two possible states: either serving a request or idle. We define the probability of entering the state α, where α is the sum of requests in service and requests in a queue, in a particular reference cycle, as S_α. The probabilities will, in the case under consideration, have to satisfy the balance equations:

$$S_0 = S_1$$
$$\frac{1}{2}(S_0 + S_1) = 1.$$

The second equation is added to normalise the probability values. As the complete reference cycle always implies two states to be passed, the sum is divided by 2. Solving these gives $S_0 = S_1 = 1$. The dependence (2) is shown in Figure 3 along with the curves obtained as a result of simulation. T_{sim} was set to 10^5 units and $S = 100$ for that case and all the cases considered below. The *mean request time* in the figures is represented on a logarithmic scale as $\log \frac{1}{\mu}$.

3 Two Request Case

This model is similar to the one considered in the previous section, except that the CPU is allowed to generate two simultaneous request to the memory. The model has three states, shown in Figure 1.

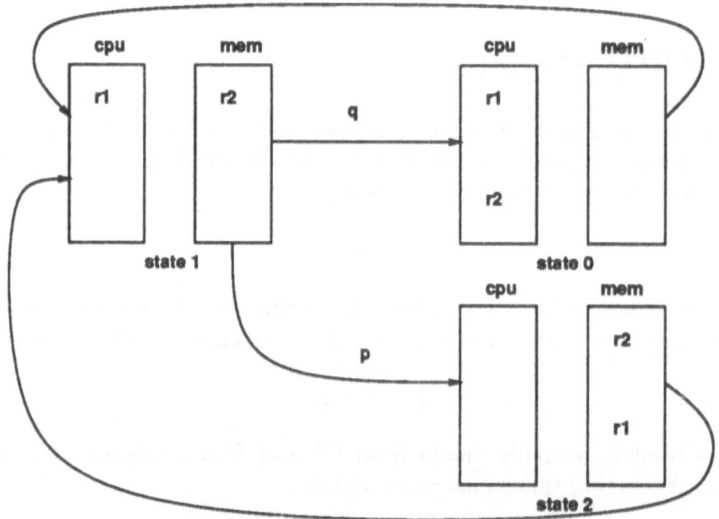

Figure 1: The states of evolution for 2 request model

The requests at the beginning of state 1 are starting to be processed by both units at the same time. The model may enter one of the two rest states (0 and 2) depending on the time when the new request will be generated. The corresponding probabilities are given by:

$$p = 1 - e^{-\mu S}$$

$$q = e^{-\mu S}.$$

Here the balance equations have the form:

$$
\begin{aligned}
S_0 &= S_1 q \\
S_1 &= S_0 + S_2 \\
S_2 &= S_1 p \\
\frac{1}{2}\sum_{i=0}^{2} S_i &= 1 \quad,
\end{aligned}
\tag{3}
$$

solving which we obtain $S_0 = q, S_1 = 1, S_2 = p$.

To calculate the average period we apply the formula:

$$< T >= \sum_{\text{all states}} S_i T_i, \tag{4}$$

where S_i and T_i are the respective probabilities and average lengths of the states. Average time spent in each of the states may be written as:

$$
\begin{aligned}
T_0 &= < t >_0^{\infty} \\
T_1 &= qS + p < t >_0^{S} \\
T_2 &= S - < t >_0^{S} \quad.
\end{aligned}
\tag{5}
$$

To calculate $< t >_0^{S}$ we have to use the probability density function recalculated to the time point S. It is known that for a NE distribution its appearance is the same for any time point [7]. Substituting (5) into (4) we obtain the final formulae:

$$
\begin{aligned}
< T > &= S + \frac{q}{\mu} \\
R &= \frac{T_{sim}}{\left(S + \frac{q}{\mu}\right)} \quad.
\end{aligned}
\tag{6}
$$

A theoretical curve obtained by this function and a simulation result are shown in Figure 2.

4 Model with Multiple Requests

To derive a balance equations for the case when n requests are allowed to be generated by the CPU we shall use *a formula of total probability* [6]

$$P(S_j) = \sum_i P(S_i)P(S_j/S_i), \tag{7}$$

where $P(S_i)$ is the probability of the event S_i and $P(S_j/S_i)$ is the conditional probability of the event S_j provided that event S_i took place. We shall consider S_i as

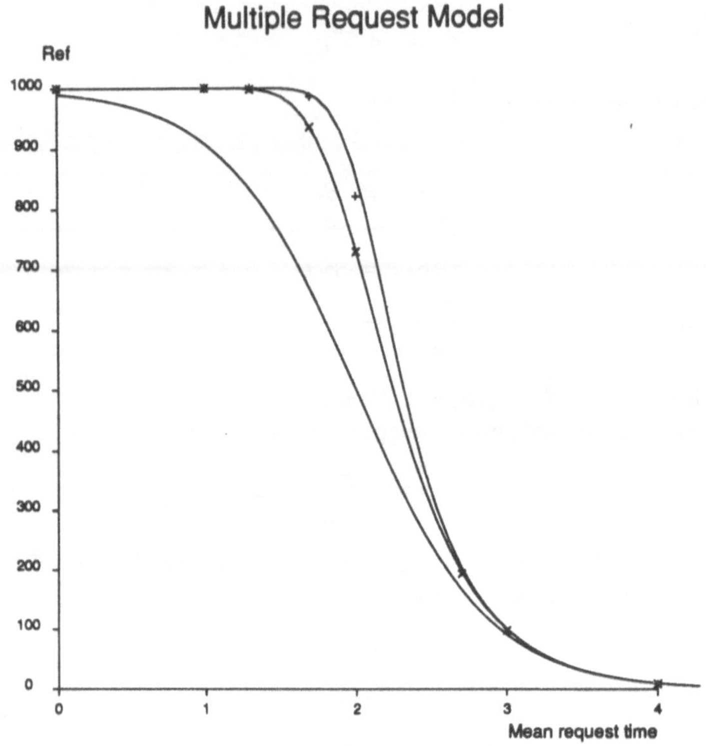

Multiple Request Model

Figure 2: The numerical data for 2 request model are × , for 3 request model are + . The theoretical curves are obtained using (6) for two request model and (16) for three request model. The left-most is a theoretical curve for one request model.

the probability to enter the state S_i in a reference cycle and the conditional probability to transfer from state S_i to state S_j as (S_i/S_j). We shall call it the transition probability. The general formula to obtain balance equations is:

$$S_j = \sum_{i \mid i \neq j} S_i(S_j/S_i). \tag{8}$$

The transition probabilities have to satisfy the equation:

$$\sum_{i \mid i \neq j} (S_i/S_j) = 1, \tag{9}$$

which means that there are no absorbing states. Using this we may write the balance equations for the multiple request model in a general form:

$$
\begin{aligned}
S_0 &= S_1(S_0/S_1) \\
S_1 &= S_0 + S_2(S_1/S_2) \\
S_k &= S_1(S_k/S_1) + ... + S_{k+1}(S_k/S_{k+1}) \\
S_{n-1} &= S_1(S_{n-1}/S_1) + ... + S_{n-2}(S_{n-1}/S_{n-2}) + S_n
\end{aligned}
\tag{10}
$$

$$S_n = S_1(S_n/S_1) + \dots + S_{n-1}(S_n/S_{n-1})$$

$$\frac{1}{2}\sum_{i=0}^{n} S_i = 1 \quad .$$

The equations for S_1 and for S_{n-1} account for the fact that the only state to be entered from the state S_0 is S_1 and from the state S_n is S_{n-1}. For other states we set the probability to enter the state S_j from the state S_i equal to the conditional probability of that event in service time interval S. It is known that the pdf for the sum of k random variables distributed according to the NE law is given by the Erlang Distribution:

$$p_k(t) = \frac{\mu^{k+1} t^k e^{-\mu t}}{k!}. \tag{11}$$

The probability that not less than k requests will be generated in time interval S is obtained by integrating (11):

$$P_k^s = 1 - e^{-\mu S} \sum_{i=0}^{k} \frac{(\mu S)^i}{i!}. \tag{12}$$

The probability to generate exactly k requests in time interval S is given by:

$$P_k^s(1 - (P_{k+1}^s/P_k^s)), \tag{13}$$

where (P_{k+1}^s/P_k^s) is a conditional probability to generate $k+1$ requests provided that k requests were generated. Substituting the value given by:

$$(P_{k+1}^s/P_k^s) = \frac{P_{k+1}^s}{P_k^s}$$

to (13), we find conditional probabilities to transfer from state S_j to S_k:

$$(S_k/S_j) = \begin{cases} P_{k-j}^s - P_{k-j+1}^s & \text{if } j < k < n \\ P_{n-j}^s & \text{if } k = n \\ q = 1 - P_{k-j}^s & \text{if } k = j-1 \end{cases} \quad .$$

Summing over k one may see that the identity (9) holds. The case $n = 2$ was considered in Section 3; for $n = 3$ setting $\rho = \mu S$ we then obtain:

$$
\begin{aligned}
S_0 &= qS_1 \\
S_1 &= S_0 + qS_2 \\
S_2 &= \rho q S_1 + S_3 \\
S_3 &= (1 - q(1 + \rho))S_1 + pS_2 \\
\frac{1}{2}\sum_{i=0}^{3} S_i &= 1,
\end{aligned}
\tag{14}
$$

solving which we may find the probability of the state S_0:

$$S_0 = \frac{2q^2}{2 - q + (1 - \rho)q^2}.$$

To calculate the average period $< T >$ we note that when the model is in a state other than S_0 the queue is always non-empty and requests are satisfied in time equal to S. The only delay is added by the idle state S_0 and is equal to $S_0 < t >_0^\infty$, which could be written as:

$$< T >= S + S_0 < t >_0^\infty . \tag{15}$$

Thus for $n = 3$ we have the formulae:

$$< T > = S + \frac{2q^2}{\mu(2 - q + (1 - \rho)q^2)},$$

$$R = \frac{T_{sim}}{S + \dfrac{2q^2}{\mu(2 - q + (1 - \rho)q^2)}}. \tag{16}$$

The theoretical curves for $n = 1, 2, 3$ and the one obtained by the simulation are shown in Figure 2. We may see that the formulae obtained give a good description for the model under consideration.

5 Asymptotic Solutions

When $q \to 0$ the first equation from (10) gives:

$$\lim_{q \to 0} S_0 = \lim_{q \to 0} qS_1 = 0,$$

which means that according to the formula (15) for the average period we have:

$$\lim_{q \to 0} < T >= S.$$

Consider the case $q \to 1$. Substituting (12) for (13) we find:

$$(S_k/S_j) = \begin{cases} \dfrac{\rho^{k-j} e^{-\rho}}{(k-j)!} & \text{if } j < k < n \\ 1 - e^{-\rho} \sum_{i=0}^{k-j} \dfrac{\rho^i}{i!} & \text{if } k = n \\ q = e^{-\rho} & \text{if } k = j + 1 \end{cases}$$

The case $q \to 1$ implies $\mu \to 0$ and $\rho \to 0$. The first of three cases obviously goes to zero. To evaluate the second case, consider the Maclaren's series for the function:

$$f(\rho) = 1 - e^{-\rho} \sum_{k=0}^{n} \frac{\rho^k}{k!}.$$

Calculating derivatives at the point 0 we find:

$$f^{(i)}(0) = \begin{cases} 0 & \text{if } 0 \le i < n \\ 1 & \text{if } i = n \end{cases},$$

so that $f(\rho)$ can be represented as:

$$f(\rho) = \frac{\rho^n}{n!} + O(\rho^{n+1}).$$

The second case also has a limit equal to zero when $q \to 1$. Considering the equation for S_n from (10) we may see that it could be represented in the form:

$$S_n = \rho F(S_1, ..., S_{n-1}),$$

where $F(S_1, ..., S_{n-1})$ is the finite function when $\rho \to 0$. Evaluating the equation from S_n to S_0 we obtain the analogous representation for the states until S_1. Substituting these into the last equation of (10) we obtain:

$$S_0 + S_1 + \rho F(S_1, ..., S_n) = 2,$$

or

$$S_0 = S_1 = 1 - O(\rho).$$

For the average period we find

$$\lim_{\mu \to 0} < T >= \lim_{\mu \to 0} S + \frac{q}{\mu} = \frac{1}{\mu},$$

which confirms the natural fact that all asymptotics are the same and equal to that for the one CPU case.

6 Shared Memory Model

The model consists of n CPUs sharing one memory unit. Each CPU generates one memory request with NE distribution and mean value $\frac{1}{\mu}$. The principal feature of this case is that the probability of entering the memory queue when the system is in the state S_0 is proportional to n. The probability density function to generate a request will be given by:

$$p(t) = \mu n e^{-\mu n t} \tag{17}$$

The general form of the balance equations is the same as (10). For the conditional probability to transfer from the state S_j to the state S_k we have:

$$(S_k/S_j) = \begin{cases} C_{n-k}^{n-j} p^{n-k} q^{k-j} & \text{if } j < k \\ \\ q^{n-j} & \text{if } j = k+1 \end{cases},$$

where C_j^i is the binomial coefficent in i and j.

Summing over k one may see that the identity (9) holds. Consider the case $n = 2$. The balance equations are identical to (3) for the two request model, and hence $S_0 = q$. The average length of state S_0 is now $\frac{1}{2\mu}$. Applying formula (15) for the average period we obtain:

$$< T >= S + \frac{q}{2\mu},$$

and for the references in term T_{sim}:

$$R = \frac{T_{sim}}{\left(S + \dfrac{q}{2\mu} \right)}.$$

In the case $n = 3$ the balance equations are:

$$
\begin{aligned}
S_0 &= q^2 S_1 \\
S_1 &= S_0 + q S_2 \\
S_2 &= 2pq S_1 + S_3 \\
S_3 &= p^2 S_1 + p S_2
\end{aligned}
\tag{18}
$$

$$
\frac{1}{2} \sum_{i=0}^{3} S_i = 1.
$$

Solving these we find:

$$
< T >= S + \frac{q^3}{3\mu(1 + \frac{1}{2}q - 2q^2 + \frac{3}{2}q^3)}.
$$

The asymptotic for S_0 when $q \to 0$ follows from:

$$
\lim_{q \to 0} S_0 = \lim_{q \to 0} q^{n-1} S_1 = 0.
$$

Applying (15) we may see that the average period is equal to the service time S. The asymptotic behaviour when $q \to 1$ could be obtained by analogy with the multi-request model considered in Section 4. For the average period we find:

$$
\lim_{q \to 1} < T >= \lim_{q \to 1} S + \frac{q^{n-1}}{n\mu} = \frac{1}{n\mu}.
$$

The asymptotic expression $S + \frac{q^{n-1}}{n\mu}$ gives a good approximation for the average period. It gives an exact value in cases $n = 1, 2$. We write the asymptotic $q \to 1$ expression for the number of serviced requests in time T_{sim} as:

$$
R = \frac{T_{sim}}{\left(S + \dfrac{q^{n-1}}{n\mu}\right)}.
\tag{19}
$$

The theoretical curves and their numerical analogues are presented in Figure 3 .

7 PRAM Model Simulation Architectures

We have considered several distributed memory architectures to simulate PRAM models. These are the Hypercube, the Torus and the Cube Connected Cycle (CCC) processor networks. A node consists of a CPU unit, local memory and a router unit. The edges are links. The CPU can generate from 1 to n memory requests to the shared memory distributed over the nodes, until it becomes idle. The requests generated have a NE distribution, and each memory request has an address associated with it. This address has a uniform distribution in the region $0 - 2^{16}$. By this we assume that a hash function should be applied to the address in a real computer system. The router unit passes a request or a message either to the local memory of the node or to the links depending on the destination address. Figure 4 reflects the ways in which the message could be transferred.

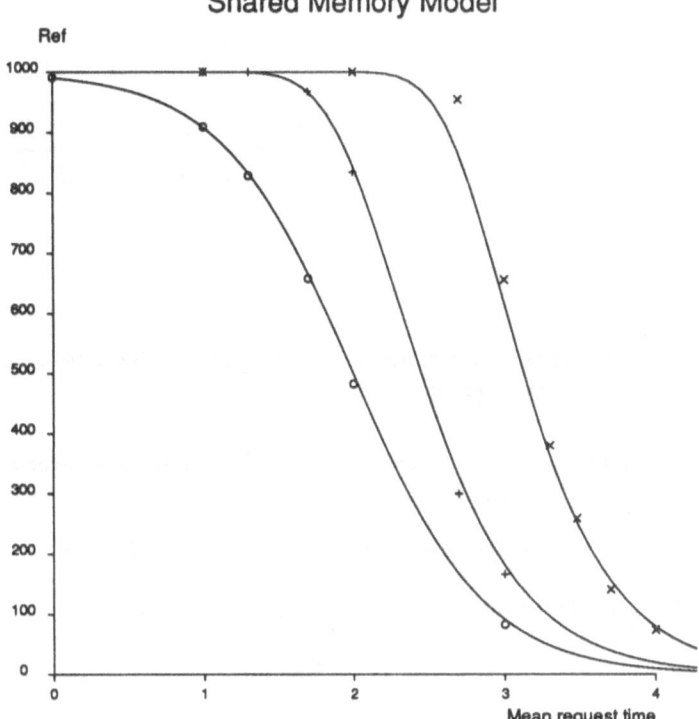

Figure 3: The numerical data for the 1 CPU model are o , for 2 CPU + , for 8 CPU × . The theoretical curves are obtained using the asymptotic expression (19).

The routing strategy was implemented as a shortest path probabilistic routing. That is, if several links are candidates to send a message along a shortest path, the link is chosen randomly among the candidates. However the probability for a link to get a message is proportional to the weight of its direction, defined as a number of different shortest length paths issuing in that direction. We expect that this routing will allow a reduction in the number of hot spots in message passing. The links were considered to be either duplex or simplex.

8 Average Message Passing Delay

To simulate the behaviour of a realistic computer system, we have to assign delays to each unit encountered as the message is transferred. We set the delays for the links and memories equal to 100 units, and the router delay is set to zero. The total time for a message to reach a destination is expressed by:

$$T_{tot} = 2 \sum_{path} (S_{link} + Q_{link}) + S_{mem} + Q_{mem}, \tag{20}$$

where the sum is taken along the path the message travels and $S_{link}, Q_{link}, S_{mem}, Q_{mem}$ are the service times and the extra time spent in a queue for each link and for

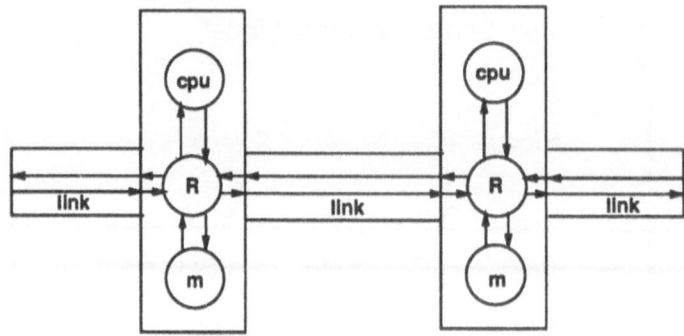

Figure 4: Communications between the functional units in the network. R denotes a router unit, M a memory module

each memory. Averaging over randomly generated requests and over all possible destinations and sources we obtain:

$$< T_{tot} >= 2d_{av}(S_{link} + < Q_{link} >) + S_{mem} + < Q_{mem} >,$$

where d_{av} is an average message passing distance, defined as:

$$d_{av} = \frac{1}{N^2} \sum_{i,j} i D_j(i),$$

where $D_j(i)$ is the number of the nodes at the distance i of the source node j. The sum over index i is taken from 1 to the maximal distance in the network, known as the diameter. Taking into account that $D_j(i)$ is equal to zero for distances more than the diameter or less than zero, we may consider that the sum is taken over all integer i. If $D_j(i)$ are identical for all nodes in the network the latter formula reduces to:

$$d_{av} = \frac{1}{N} \sum_i i D(i).$$

For an n-dimensional Hypercube it has been shown [8] that $D(k) = C_k^n$ thus for d_{av} we obtain:

$$d_{av} = \frac{1}{N} \sum_i i C_i^n = n/2,$$

where $N = 2^n$. The table below gives the numerical values of average message distance for Hypercube and CCC.

Dimension	3	4	5	6
CCC		3.83	4.625	5.95	7.541
Hypercube	1.5	2	2.5	3

9 Analysis of the Simulation Results

As a result of simulation we found the dependency of the number of references, made to the memory, on the requests distributed according to the NE law for different mean values. The simulation was carried out for Hypercubes up to D=7, for CCCs with *the arity* = 3 up to D=5 and for Torii (p-ary 2 cubes) up to 8-ary 2 cube, using duplex and simplex models of the links and for 3 different request models for each CPU. Figure 5 shows the curves for a 3D Hypercube and a 3D CCC.

One may see that, depending on the number of requests allowed per CPU, the saturation limit is different. For single-request models the average waiting time spent in link and memory queues was observed to be less than 10% of the service time, for all mean request values. If we ignore $< Q_{link} >$ and $< Q_{mem} >$ in (20), we obtain a formula that gives a satisfactory performance prediction in terms of references to the memory.

$$R = \frac{T_{sim}}{\left(2d_{av}S_{link} + S_{mem} + \frac{1}{\mu}\right)}. \tag{21}$$

Comparing the curves we may notice that:

- The asymptotic behaviour for $\frac{1}{\mu} \to \infty$ is identical and equal to μ.

- The Hypercube network has at least twice as many requests satisfied compared to the CCC for single and multi-request models.

- The curves for multi-request models have an inflection point that sets a boundary between saturated and non saturated networks.

- The Hypercube network with 16 requests per CPU is very close to the theoretical limit equal to 1000 references for 10^5 units simulation time and 100 units memory access time.

The total number of references can be obtained as a product of the number of references per CPU and the number of CPU in the network. This allows us to obtain a network performance in terms of the number of references to the memory.

In Figure 6 the results of numerical simulation for 6D Hypercube network are represented. The models investigated had either duplex or simplex links. Comparing the values of the references observed, we see that for single-request model they are very close. This can be explained by the small values of $< Q_{link} >$ and $< Q_{mem} >$ in (20) for that case. The formula (21) can be applied with less accuracy for the simplex link model too. For multi-request models the difference in the asymptotic $\frac{1}{\mu} \to 0$ goes to two when $\#Requests \to \infty$. In fact it is close to two for $\#Requests = 16$ in the case under consideration.

The analogous simulation results for the 8-ary 2 Cube are shown in Figure 7. The ratio of the maximal number of the references observed for 6D Hypercube to that for 3D Hypercubes is $\frac{820}{930} \simeq 0.88$, the analogous ratio for 8-ary and 3-ary 2 cubes is $\frac{457}{984} \simeq 0.46$. The number of nodes in the ratios considered is the same for 6D Hypercube and 8-ary 2 cube, equal to 64, and is in proportion $\frac{8}{9}$ for 3D Hypercube to 3-ary 2 cube. The ratios show the decrement rate in the references observed per CPU with the growth of the number of CPUs in the network.

10 Acknowlegments

I am very grateful to the EPCC for inviting me to work there. My special thanks are to G.V.Wilson for his ideas on how to simulate the computer architectures. I am also very thankful to R.J.Pooley, M.A.Smith and M.D.Brown for their help.

References

[1] S.G. Akl, "The Design and Analysis of Parallel Algorithms," Prentice-Hall International, Inc., 1989.

[2] L.G. Valiant, "General Purpose Parallel Architectures," Handbook of Theoretical Computer Science, Elsevier Science Publishers B.V.,1990.

[3] P.G. Harrison, N.M. Patel, "The Representation of Multistage Interconnection Networks in Queueing Models of Parallel Systems," Journal of the ACM, 37, no 4, pp. 863-898, Oct. 1990.

[4] R.J. Pooley, "An Introduction to programming in Simula," Blackwell Scientific Publications, 1987.

[5] G.M. Birtwistle, "Discrete Event Modelling on Simula," Macmillan Education Ltd., 1979.

[6] B.V.Gnedenko, "The Theory of Probability," Chelsea Publishing Company, New York, 1966.

[7] E. Page, "Queueing theory in OR," London Butterworths, 1972.

[8] L.N. Bhuyan, D.P. Agrwal, "Generalised Hypercube and Hyperbus Structures for a Computer Network", IEEE Trans.Comput. vol C-33, pp. 323-332, Apr. 1984.

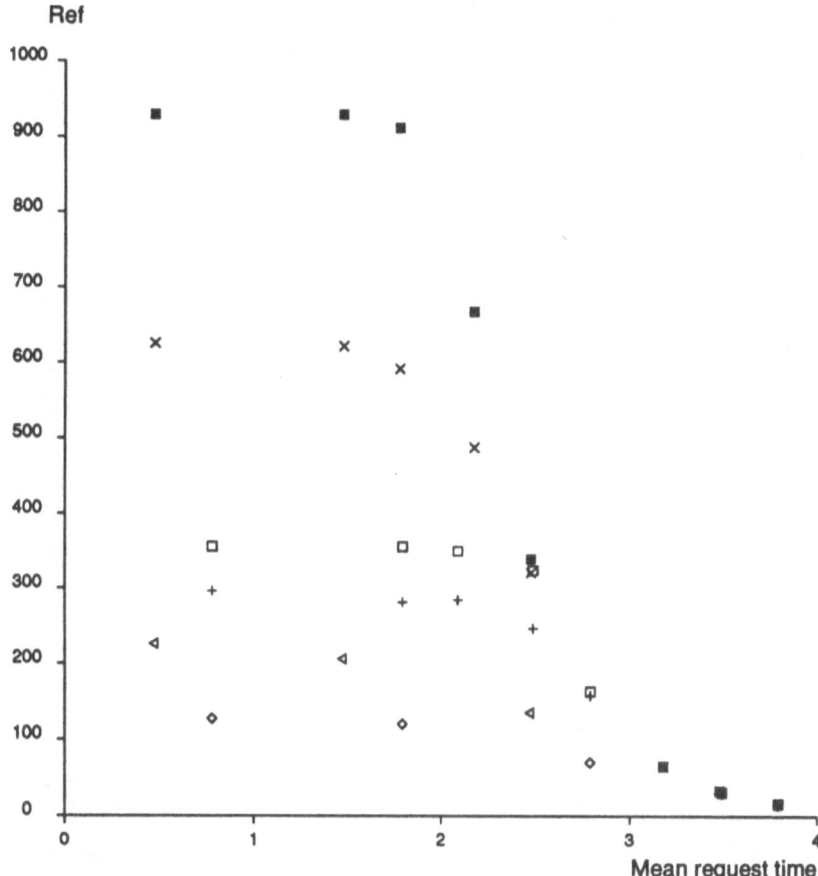

Figure 5: The number of references observed for 3D Hypercybe 16 request model are dark squares, 4 request–crosses, 1 request–triangles, for CCC 16 request model are white squares, 4 request–pluses, 1 request–rhombuses.

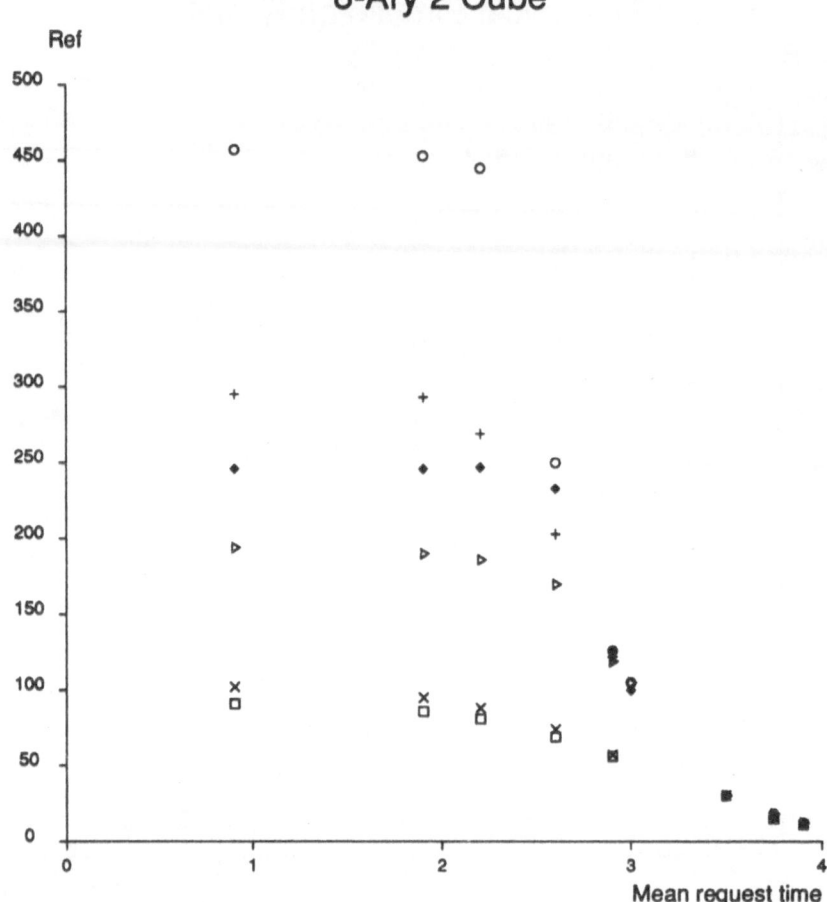

Figure 6: The results of simulation for 6D Hypercube. Crosses, circles and pluses are the marks for duplex link models with 16, 4 and 1 request generated per CPU. Squares, triangles and rhombuses are marks for simplex link models with 16, 4 and 1 request per CPU.

8-Ary 2 Cube

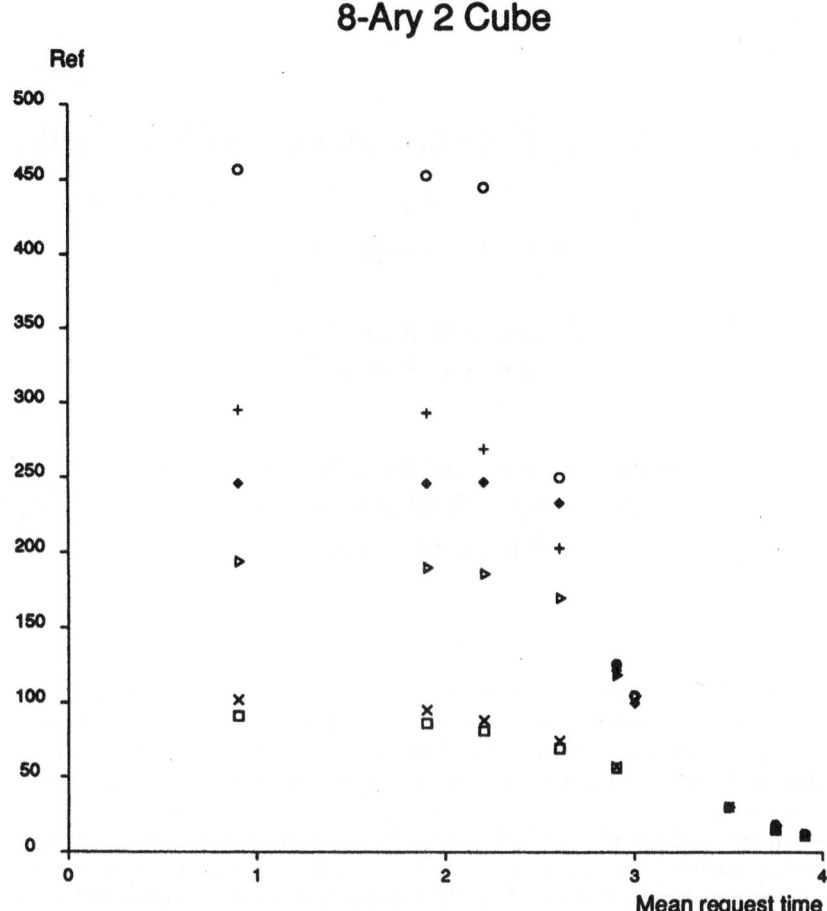

Figure 7: The results of simulation for 8-ary 2 cube. Circles, pluses and crosses are the marks for duplex link models with 16, 4 and 1 request generated per CPU. Rhombuses, triangles and squares are marks for simplex link models with 16, 4 and 1 request per CPU.

General Queueing Network Models with Variable Concurrent Programming Structures and Synchronisation[*]

Demetres D Kouvatsos
Andreas Skliros

Computer Systems Modelling Research Group
University of Bradford, Bradford
United Kingdom

Abstract

A significant computational gain can be achieved through concurrent programming by exploiting parallel and distributed processing capabilities of current computer system architectures. The performance evaluation and quantitative analysis of such systems have become important due to the multiplicity of the component parts and the complexity of their functioning. In this paper an approximate method is developed for the analysis of general queueing network models of computer systems with variable concurrency and synchronisation structures. It is based on the maximum entropy algorithm and the notion of surrogate delays. Numerical examples illustrate the capability of the proposed algorithm in comparison to simulation.

1 Introduction

The analysis of distributed systems has attracted the interest of scientific research due to the availability of inexpensive, high performance and reliable digital computers. Increasing computational throughput is achieved by exploiting the parallel processing capabilities of the system architecture through the inherent structural parallelism of concurrent programming languages like ADA, OCCAM, Concurrent Pascal and Modula. Thus, the performance evaluation and prediction of such systems are both vital and urgent issues.

The simultaneous occurrence of various events is a frequent phenomenon in real-time embedded environments or operating systems, and thus, the need for programming languages with concurrent instruction execution becomes indisputable [1,2]. A concurrent program is a set of ordinary sequential subprograms, called tasks, that are executed in abstract parallelism. Moreover, the tasks are processed concurrently

---

[*]This work is sponsored by the Science and Engineering Research Council (SERC), UK, under grant GR/F29271

and independently of one another in either centralised or distributed computer systems.

Parallel processes can coordinate under various hardware environments. They either communicate by exchanging values and messages or get synchronised by transmitting the status of a process or a system. Two approaches have mainly been used to implement the interprocess interactions. The first one is based on the use of shared variables and is utilised in synchronisation mechanisms like busy-waiting, semaphores, conditional critical regions, monitors and path expressions. The second approach is based on message passing and is implemented in the communication channel and the remote procedure call mechanisms.

Real-time embedded computer systems in aircraft, radar installations and power plants widely use concurrent programming. Queueing network models (QNMs) have been employed to study the performance of such systems. However, their analysis is not a trivial task due to the complex functioning of the concurrent processes along with their synchronisation requirements and the unpredictable pattern of their service demands. Exact solutions and simulations have been proved computationally intractable and prohibitively costly. Approximate analytic techniques appear more consistent and promising.

Earlier research on models with program concurrency was focused on the overlap of buffering systems or between CPU and I/O units [3–6]. The concurrent tasks are tightly coupled allowed to visit only one resource before getting synchronised. Herzog et al [7] modelled a system with job concurrency assigning each parallel process to its own processor. Bard [8] described an approximation for a class of concurrent programs based on the mean value analysis algorithm [9]. Heidelberger and Trivedi [10] presented two Markovian models with fixed concurrency based on the notions of decomposition and complementary delays, respectively, allowing the tasks to access many resources before synchronisation. Other research work [11, 12] although interesting tends to be of limited applicability and rather costly to extend into more complex cases.

Although earlier models have captured some of the main features of concurrent programming their general applicability is restricted by the following modelling assumptions:

(a) Markovian chains are used with exponentially distributed service times having a common mean for all classes of tasks;

(b) Service disciplines at the queueing stations must agree with the conditions of the product-form networks [9];

(c) The structures of concurrency and synchronisation do not allow variable number of concurrent tasks, nested concurrency or any type of synchronisation schemes.

A recently proposed algorithm [13] based on entropy maximisation and hierarchical decomposition attempts to relax all the above mentioned assumptions. This algorithm has been shown to be very credible and has also been employed to examine the effect of the service time variability on the performance gain due to multi tasking and multiprocessing [14]. However, it is rather costly in modelling QNMs with a large number of queueing centres and classes of tasks under nested concurrent forms.

In this paper an approximate algorithm is proposed for the analysis of QNMs with variable concurrent structures, flexible synchronisation schemes, generally distributed service times of the tasks and stations with mixed service disciplines (in-

58

cluding priority schemes). It is based on the maximum entropy algorithm [15] and surrogate delays [16].

Section 2 describes the technique to model fixed concurrency and synchronisation of programs. The so-called ME-DELAY approximate algorithm for analysing QNMs with fixed concurrency is presented in Section 3. A variable concurrent and synchronisation structure is modelled in Section 4 and an approximate solution is proposed. The credibility of the proposed algorithms is demonstrated in Section 5. Conclusions and directions for future research follow in Section 6.

2   Fixed secondary concurrent level model (FS Model)

2.1   Background

Programs with concurrency and synchronisation schemes executing in local or distributed computer systems include a large amount of statements and data. For modelling purposes, these programs are filtered in order to extract only the information regarding the issues of concurrency and synchronisation. This information is illustrated in a task graph that serves as input to the modelling process. A program task graph shows transparently the execution life of the program, the cycle of the parallel processes and the synchronisation points (called rendezvous).

In Figure 1 a program (pseudocode) in ADA [17] is depicted with simple concurrent and synchronisation structures. At a time instant t_0 the entire job PROGRAM_FS initiates execution. The PRIM_1 procedure, consisting of sequential statements, receives service until time instant t_1 when it completes. Since only one instruction belonging to PROGRAM_FS is executed at a time, it is said that a primary task is in the active stage during the time interval (t_0, t_1) which is called the first cycle.

At its end, the procedure PARALLEL_1 begins execution. It consists of two parallel processes SEC_11 and SEC_12, each of which includes sequential code. Despite this, more than one instruction belonging to the same program can execute at a time and the concurrent tasks are called secondary tasks or siblings. During the (t_1, t_2) time interval both secondary tasks are active receiving service concurrently and independently of one another. At time t_2 SEC_11 completes execution and starts to experience a delay equal to interval, (t_2, t_3) as long as its sibling SEC_12 remains active. At time t_3 both the secondary tasks synchronise and pass the control of the program to the next procedure. This type of synchronisation can model asynchronous or buffered message-passing synchronisation schemes. The interval (t_1, t_3) forms another cycle. During the third cycle (t_3, t_4) the primary task PRIM_2 is active.

The fourth cycle (t_4, t_6) includes the parallel execution of the first parts of the tasks SEC_21 and SEC_22. Inside the code of these two tasks there is a synchronisation block that prohibits the execution of the rest of the code, until both tasks reach this block and execute the synchronisation commands. The SEC_22 task reaches this synchronisation point at t_5 and is delayed until its sibling arrives at the rendezvous at time t_6. This type of synchronisation is referred to as synchronous. Following the message exchange, both secondary tasks resume execution, as SEC_31 and SEC_32 respectively, and after completing the whole task body, they synchronise again at

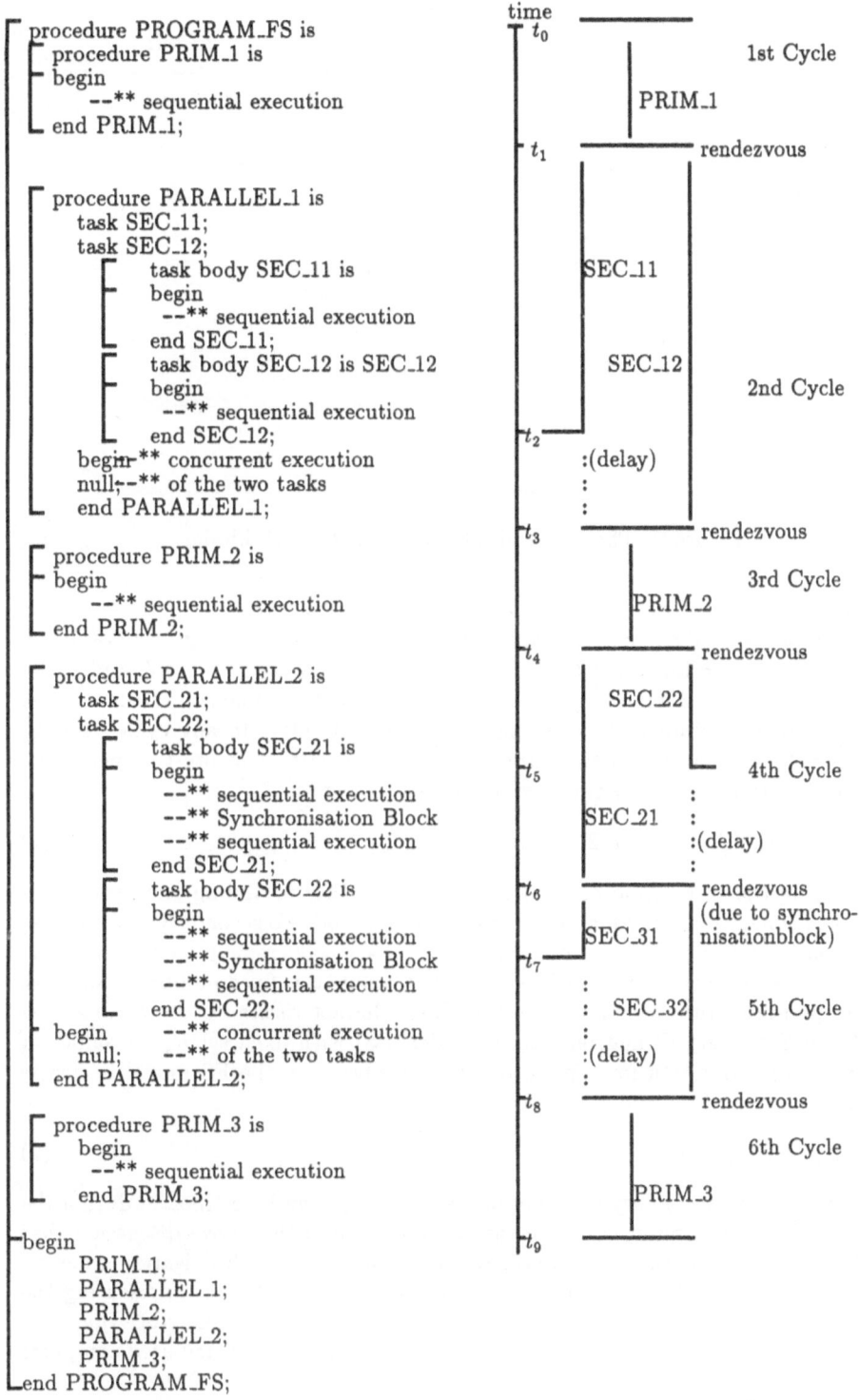

Figure 1: Program in ADA with fixed concurrency and the task graph (FS)

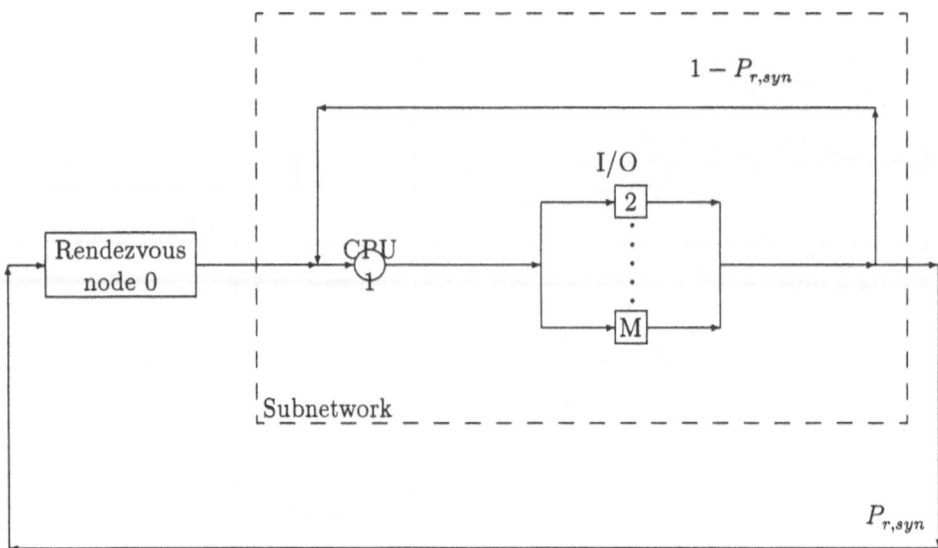

Figure 2: The Centralised Queueing Network Model

time t_8 and pass control to the next procedure for a new cycle.

In a more general context, a quantitative analysis of the resulting task graph can be carried out as follows. Let C be the number of the cycles in a graph and C_r be the number of cycles including r active tasks, $r = 1, 2, \ldots, \max(s)$, where $\max(s)$ is the maximum number of tasks that can be found active in any cycle. Let V_s denoting the secondary splitting probabilities vector with elements $V_{s,r}$ measuring the probability to find r tasks active in a cycle i.e.,

$$V_{s,r} = C_r/C, \qquad r = 1, 2 \ldots, \max(s) \tag{1}$$

Quantifying the task graph of Figure 1, the probability of finding one task active (i.e., the primary task) is equal to the probability to find two secondary tasks active, both equal to 0.5. Since the number of the secondary tasks remains constant, let R_s denote the secondary concurrent level (scl) that counts this fixed number of siblings. Hence, programs that accommodate the above characteristics can be analysed by the Fixed Secondary Concurrent Level Model (FS) with fixed scl R_s. A sufficient and necessary condition for a program to be analysed by the FS model is clearly given by

$$V_{s,1} = V_{s,R_s} = 0.5 \tag{2}$$

Furthermore, the secondary splitting probabilities $V_{s,r}$ can be estimated during the monitoring of a computer environment. It follows from the above discussion that the FS model can capture the described concurrency and synchronisation schemes. The next step is to fit this model within the context of a QNM representing the hardware configuration of interest under mixed service disciplines. Note that programs modelled by the FS can execute either in centralised or distributed computer systems.

Figure 2 illustrates the QNM associated with the task graph of Figure 1 and is comprised of two parts. The first part is the rendezvous node (labelled 0), a delay type queueing station under Infinite-Service (IS) discipline, where the synchronisation of the tasks takes place. The second part includes a subnetwork consisting of M single server infinite capacity queueing stations with mixed service disciplines including non-priority rules such as First-Come First-Served (FCFS), Last-Come-First-Service with (LCFS-PR) or without (LCFS) preemption, Processor Sharing (PS), or priority schemes like Head-of-the-Line (HOL) and Preemptive-Resume (PR).

Note that the use of the central server model is not a restriction, because each node of a distributed network can represent a local centralised system that is connected to other local systems via communication lines.

According to the FS model each program can be viewed as partitioned into one primary task (parent process) and R_s secondary tasks (child processes), each with different service requirements. Therefore, each task belongs to a different class and there are $(R_s + 1)$ classes of tasks in the network, labelled with 0 the primary class and r, $r = 1, \ldots, R_s$, the classes of the secondary tasks. The queueing network is represented by a closed model having N whole programs, which implies that the multiprogramming level is $n = (n_0, n_1, \ldots, n_{R_s}) = (N, N, \ldots, N)$.

Each service station i, $i = 1, \ldots, M$, provides a generally distributed service per class r, $r = 0, 1, \ldots, R_s$, with rate $\mu_{i,r}$ and squared coefficient of variation (scv) $C^2_{s\ i,r}$. The tasks visit the nodes according to routing probabilities $(P_{r,i,j})$ (first-order Markov Chain) such that $P_{r,i,j}$ is the probability a task of class r to visit node j following departure from node i, $i, j = 0, 1, \ldots, M$.

2.2 The FS queueing network model

The FS model within the context of a QNM is described as follows. At an initial state all the primary and secondary tasks are in the rendezvous centre. The first cycle of each program starts either with the primary task being active with probability $V_{s,1}$ or with all R secondary tasks becoming active with probability V_{s,R_s}, where a Bernoulli trial determines the future active tasks.

Let us assume that the primary task initiates the first cycle. It departs from the rendezvous node and enters the subnetwork. As long as it visits various centres receiving service, it is considered to be in the active stage. Upon completion, it exits from the subnetwork with probability $P_{0,syn}$ and arrives at the rendezvous node, finishing the cycle. Note that the siblings of an active primary task experience delay at the rendezvous centre, as long as it remains active.

The Bernoulli experiment is repeated and suppose the secondary tasks have been chosen to become active in the second cycle. Hence, all the child processes, belonging to the parent process that was active in the previous cycle, enter the subnetwork. They execute concurrently and independently of one another at the queueing centres of the network, except for the waiting effect where two or more siblings compete for the same resource.

A secondary task of class r, $r = 1, \ldots, R_s$ can depart from the subnetwork (terminating its active stage) and enter the rendezvous node with probability $P_{r,syn}$. This probability is a tuning parameter and measures the frequency of the synchronisation demands. High values imply frequent synchronisation resulting to increased synchronisation overhead and lower computational gain. The value of the synchronisation probability, $P_{r,syn}$, imposes no restriction to the applicability of the FS model, in contrast to earlier research assuming either frequent or loose synchronisation. Note that each secondary task requires synchronisation because it either completes execution or reaches a synchronisation block in its code body that forces the particular task to suspend execution and be delayed until it gets synchronised with the rest of its siblings.

Upon the secondary task entering the rendezvous node it checks if all of its $(R_s - 1)$ siblings have been waiting for its arrival. If this is the case, all siblings get synchronised and the cycle is completed. Otherwise, it initiates a delay period (wait stage) until all its siblings arrive at the rendezvous node. Following their synchronisation (merging) another Bernoulli experiment takes place to determine the next cycle.

This modelling mechanism, utilising the secondary splitting probabilities, provides a flexible structure for the FS model which captures a wide range of synchronisation schemes, under a general distributional pattern of the tasks' service demand, while it is independent of the hardware configuration (centralised or distributed network). However, its applicability is limited by the fixed concurrent structure, a feature that is relaxed in the new model introduced in Section 4.

3 The ME-DELAY approximate algorithm

The effect of the concurrent programming structures and synchronisation can be captured by estimating the mean and scv of the delay time at the rendezvous node. This section describes the ME-DELAY algorithm, which iteratively applies the UME algorithm [5, 8] in conjunction with the method of surrogate delays [16] in order to approximate the above parameters and produce the final performance measures of the entire FS QNM.

3.1 The random delay times

The tasks upon entering the rendezvous node experience a delay determined by the rules governing the synchronisation schemes. The durations of these delays are not known in advance and they have to be estimated analytically. Let D_r and L_r denote the random variables representing the delay imposed to the task of class r and the active time (being in the subnetwork) of this particular task, $r = 0, 1, \ldots, R_s$, respectively.

For the FS model $V_{s,1} = V_{s,R_s} = 0.5$ and, therefore, the Bernoulli trials on average will determine in one cycle the primary task to be active and be followed by the secondary ones at the next cycle. Hence, a primary task will be delayed as long as the duration of the cycle with the active siblings which is equal to the response time of the longest secondary task. It is, therefore, implied that

$$D_r = \max\{L_1, L_2, \ldots, L_R\} \tag{3}$$

Each secondary task is delayed at the rendezvous centre as long as either its parent or its siblings are active. Thus, the overall delay, D_r, is clearly given by

$$D_r = L_0 + \max\{L_1, L_2, \ldots, L_R\} - L_r = L_0 + D_0 - L_r, \tag{4}$$

where $r = 1, 2, \ldots, R_s$.

Hence, it is essential to estimate the distribution of the delays D_r, $r = 0, 1, .., R_s$, or at least the first two moments, mean $E(D_r)$ and scv $C^2(D_r)$ through formulae (3) and (4). This requires the prior estimation of the first two moments of the active times, namely $E(L_r)$ and $C^2(L_r)$. The active time of a task is actually its response time in the subnetwork and can be viewed as the sum of the residence times $L_{i,r}$, $r = 0, 1, \ldots, R_s$ and $i = 1, \ldots, M$. Since delays are initially unknown, an iterative algorithm seems to be a suitable technique in order to manipulate the active and delay times per iteration until convergence.

3.2 The UME algorithm

The UME algorithm can be applied to utilise a product-form approximation for an extended class of general closed QNMs under mixed service disciplines [17]. The algorithm is applied in two stages.

First Stage: This involves an initial decomposition of the closed QNM into individual infinite capacity G/G/1 queues (referred to as "pseudo-open network" [15]) satisfying constraints on the flow conservation and population conservation. The Generalised Exponential (GE) distribution of the form:

$$F(t) = 1 - \frac{2}{1 + C^2} \exp \left\{ -\frac{2\tau t}{1 + C^2} \right\} , \quad t \geq 0$$

where $1/\tau, C^2$ are the mean and scv, respectively, is used to approximate general distributions of interest with given first two moments. In this way Lagrange multipliers corresponding to the marginal constraints of utilisation, mean queue length and idle state probability can be determined.

Second Stage: This stage incorporates into the ME product-form solution of closed QNMs the estimates of the Lagrange multipliers of the first stage and applies a convolution based technique to calculate the maximum entropy joint state probability iteratively until the job flow balance equations, as applied to the original closed QNM, are satisfied.

More details of the UME algorithm can be found in [15, 18].

3.3 The ME-DELAY algorithm

The general steps of the ME-DELAY algorithm can be described as follows.

Step 1: Initialise $E(D_r)$ and $C^2(D_r)$

Step 2: Repeat until convergence in $E(D_r)$ and $C^2(D_r)$.

> Step 2.1: Apply the first stage of the UME algorithm, for solving the general pseudo-open QNM and estimate the measures $E(L_{i,r})$ and $C^2(L_{i,r})$

> Step 2.2: Estimate the active time measure $E(L_r)$ and $C^2(L_r)$.

> Step 2.3: Estimate the delay time measures $E(D_r)$ and $C^2(D_r)$

Step 3: Apply the second stage of the UME algorithm in order to flow-balance the general closed QNM and produce the final performance measures.

In a more detailed fashion, steps 2 and 3 are presented as follows.

Step 2.1: Apply the first stage of the UME algorithm to solve the pseudo-open network with $M+1$ queueing centres, R_s+1 classes and $N*(R_s+1)$ independent tasks. Calculate the mean $\lambda_{i,r}$ and the scv $C^2_{a\,i,r}$ of the interarrival process per class and solve each queueing station as a GE/GE/1 multiple class queue with given mean service rate $\mu_{i,r}$ and scv $C^2_{s\,i,r}$ per class r. Estimate the associated Lagrange multipliers and determine the GE/GE/1 mean $E(L_{i,r})$ and scv $C^2(L_{i,r})$ of the response them per class $r, r = 0, 1, \ldots, R_s$, at centre $i, i = 1, \ldots, M$, approximated by [19]

$$E(L_{i,r}) = \frac{\tau_{i,r} * (1 - \rho_{i,r} * \sigma_{i,r}) + \rho_{i,r} * \sigma_{i,r}}{\sigma_{i,r} * \tau_{i,r} * \mu_{i,r} * (1 - \rho_{i,r})} \tag{5}$$

$$C^2(L_{i,r}) = \frac{2 * \tau_{i,r} * (1 - \sigma_{i,r}) + 2 * \sigma_{i,r}}{\tau_{i,r} * (1 - \rho_{i,r} * \sigma_{i,r}) + \rho_{i,r} * \sigma_{i,r}} - 1 \tag{6}$$

where $\rho_{i,r} = \lambda_{i,r}/\mu_{i,r}$, $\tau_{i,r} = 2/(1 + C^2_{s\,i,r})$ and $\sigma_{i,r} = 2/(1 + C^2_{a\,i,r})$.

Step 2.2: The active time of each task is the sum of the response times of the centre it visits. Let L^f_r be the time period during which a task is active in the subnetwork of Figure 2, assuming there is no feedback (i.e., the system behaves in a feedforward fashion). Clearly, this can be expressed by:

$$L^f_r = L_{1,r} + \sum_{j=2}^{M} P_{1,j,r} * L_{j,r}, \tag{7}$$

with mean, variance and scv, respectively, given by

$$E(L^f_r) = E(L_{1,r}) + \sum_{j=2}^{M} P_{1,j,r} * E(L_{jr}) \tag{8}$$

$$Var(L^f_r) = Var(L_{1,r}) + \sum_{j=2}^{M} P_{1,j,r} * Var(L_{j,r}), \tag{9}$$

$$C^2(L^f_r) = Var(L^f_r)/E(L^f_r)^2. \tag{10}$$

However, queueing centres of the subnetwork could be visited several times depending on the feedback probability $(l - P_{r,syn})$. Thus, the overall mean and scv of the active items are given, respectively, by:

$$E(L_r) = E(L_r^f)/P_{r,syn}, \tag{11}$$

$$C^2(L_r) = (1 - P_{r,syn}) + P_{r,syn} * C^2(L_r^f). \tag{12}$$

Step 2.3: The mean $E(D_0)$ and variance $Var(D_0)$ of the delay time of a primary task can be estimated by assuming: (i) the general distribution of response time L_r is approximated by a GE type distribution with parameters given by (11) and (12); (ii) active times per class are independent of each other.

To this end, it can be verified, using (3), that the mean, variance and scv of the mean value of GE-type random variables $\{L_1, L_2, \ldots, L_{R_s}\}$ are given, respectively, by the following expressions:

$$E(D_0) = \sum_{r_1=1}^{R_s} \frac{1}{\ell_{r_1}} - \sum_{r_1 < r_2} \frac{W_{r_1} * W_{r_2}}{(W_{r_1} * \ell_{r_1}) + (W_{r_2} * \ell_{r_2})} + \cdots +$$
$$(-1)^{R_s - 1} \sum_{r_1 < \cdots < r_{R_s}} \frac{W_{r_1} * \cdots * W_{r_{R_s}}}{(W_{r_1} * \ell_{r_1}) + \cdots + (W_{r_2} * \ell_{r_{R_s}})}, \tag{13}$$

$$Var(D_0) = \sum_{r_1=1}^{R_s} \frac{2}{W_{r_1} * \ell_{r_1}^2} - \sum_{r_1 < r_2} \frac{2 * W_{r_1} * W_{r_2}}{((W_{r_1} * \ell_{r_1}) + (W_{r_2} * \ell_{r_2}))^2} + \cdots +$$
$$(-1)^{R_s - 1} \sum_{r_1 < \cdots < r_{R_s}} \frac{2 * W_{r_1} * \cdots * W_{r_{R_s}}}{((W_{r_1} * \ell_{r_1}) + \cdots + (W_{r_{R_s}} * \ell_{r_{R_s}}))^2} - E(D_0)^2, \tag{14}$$

$$C^2(D_0) = Var(D_0)/E(D_0)^2 \tag{15}$$

where $\ell_r = 1/E(L_r)$ and $W_r = 2/(1 + C^2(L_r))$ for $r = 1, \ldots, R_s$.

Moreover, mean, variance and scv of the delay times of these secondary tasks of class r, are given according to (3.2), by the following formulae:

$$E(D_r) = E(L_0) + E(D_0) - E(L_r), \tag{16}$$

$$Var(D_r) = Var(L_0) + Var(D_0) + Var(L_r) \tag{17}$$

$$C^2(D_r) = Var(D_r)/E(D_r)^2 \tag{18}$$

Step 3: Applying the second stage of the UME algorithm in order to obtain the final performance measures of the QNM with FS type concurrency and synchronisation, namely throughput $X_{i,r}$, utilisation $U_{i,r}$ and mean queue length $\langle n \rangle_{i,r}$ of the task belonging to class $r, r = 0, 1, \ldots, R_s$, at queueing centres $i, i = 0, 1, \ldots, M$.

4 Variable secondary concurrent level (VS) model

This section presents the VS model for the analysis of programs with variable concurrent structure. It maintains the advantages of the FS model capturing synchronisation schemes and service time variability while it corrects the drawbacks imposed by the fixed concurrent structure. Figure 3 displays a program in ADA and the corresponding task graph. The number of tasks being active in a cycle follows a random pattern that is governed by the secondary splitting probabilities which can take any value. For example, the corresponding V_s vector for the task graph of Figure 3 is given by $V_s = (V_{s,1}, V_{s,2}, V_{s,3}, V_{s,4}) = (0.4, 0.2, 0.2, 0.2)$. Clearly, the FS model is not appropriate to represent such concurrent structure. The VS model can be characterised by defining R_s' as the average secondary concurrent level, estimated through the formula

$$R_s' = \sum_{r=1}^{\max(s)} r V_{s,r} \qquad (19)$$

The notion of an average scl is characteristic of the FS model too. Applying (19), the following scale with the corresponding average and fixed scls is produced:

FIXED scl	(R_s)	2	3	4	5	
AVERAGE scl	(R_s')	1.5	2	2.5	3	...

Hence, the estimation of the average scl of a model can be ranged within this scale. It will lie between two values that correspond to FS models. These models are called the lower and the upper neighbouring FS models of the VS model. Their average and fixed scls are $\left\{R_s^{L'}, R_s^{U'}\right\}$ and $\left\{R_s^L, R_s^U\right\}$, respectively. The neighbouring fixed scl can be estimated by the following expressions

$$R_s^L = trunc\left((R_s^{L'}/0.5) - 1\right), \qquad (20)$$

$$R_s^U = round\left((R_s^{L'}/0.5) - 1\right). \qquad (21)$$

Therefore, it is reasonable to assume that the performance of the VS model can be approximated by averaging the performance of the two neighbouring FS models. Furthermore, when the average scl of the VS model lies closer to the average scl of one of the FS neighbouring models, it is implied that the performance of the VS model must be closer to the performance of that particular FS model. Following this argument application of a weighted average is imposed. Let Z_L and Z_U be the weights of the lower and upper neighbouring FS models, respectively, given by

$$Z_U = \frac{(R_s^{U'} - R_s^{L'}) - (R_s^{U'} - R_s')}{(R_s^{U'} - R_s^{L'})} = \frac{0.5 - (R_s^{U'} - R_s')}{0.5}, \qquad (22)$$

$$Z_L = 1 - Z_U \qquad (23)$$

For example, the task graph of Figure 3 has average scl $R_s' = 2.2$, neighbouring FS models with fixed scls $R_s^L = 3$ and $R_s^U = 4$ and weights $Z_L = 0.6$, $Z_U = 0.4$ This means that the performance measures ('performance') of VS satisfies generally the following relation:

67

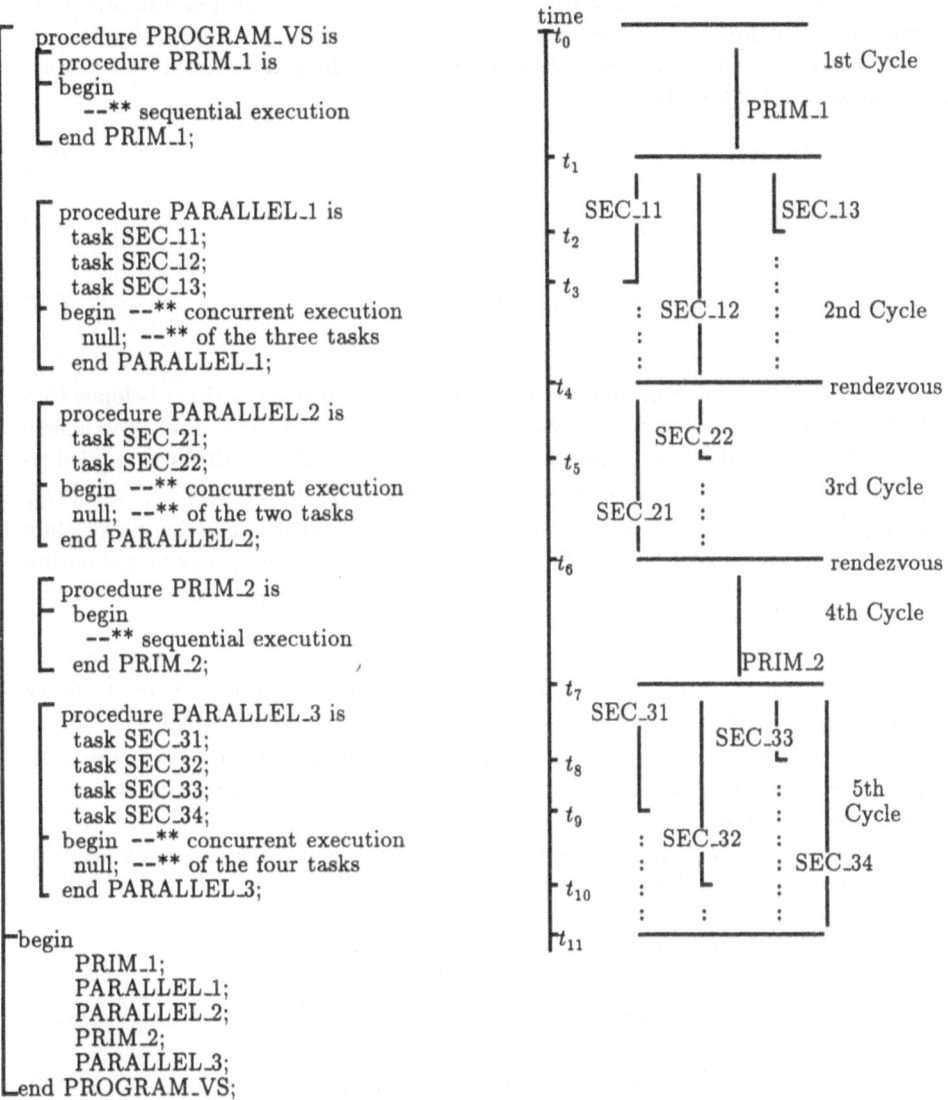

Figure 3: Program in ADA with variable concurrency & the task graph (VS)

$Performance(VS) = 0.6*Performance(FS(R_s^L = 3))+0.4*Performance(FS(R_s^U = 4))$.

An implementation of the above approximation per class of tasks is presented below. Let A_s denote the active secondary tasks' probability vector with elements $A_{s,r}$, $r = 0, 1, \ldots, \max(s)$ measuring the probability to find in a cycle a task belonging to class r, estimated by:

$$A_{s,0} = 1 - A_{s,2} = V_{s,1}, \tag{24}$$

$$A_{s,1} = A_{s,2} \tag{25}$$

$$A_{s,r} = \sum_{i=r}^{\max(S)} V_{s,i}, \; for \; r = 2, \ldots, \max(S) \tag{26}$$

It follows that when a cycle has r ($r > 1$) active tasks each of them belongs to a different class, i.e., one belongs to class 1, one to class 2 and so on. Since the neighbouring FS models have been determined, the ME-DELAY algorithm is applied to approximate the performance measurements of these two models. Let $X_{i,r}^L, U_{i,r}^L, \langle n \rangle_{i,r}^L$ and $X_{i,r}^U, U_{i,r}^U, \langle n \rangle_{i,r}^U$ be the throughput, utilisation and mean queue length of class $r, r = 0, 1, \ldots, R_s$ at centre $i, i = 0, 1, \ldots, M$ of the lower and upper neighbouring FS models respectively and $X_{i,r}', U_{i,r}', \langle n \rangle_{i,r}'$ the corresponding performance statistics of the VS model.

In particular, let $X_{i,0}^L$ be the throughput of the primary tasks at the rendezvous node of the lower FS model where the secondary active probability for the primary $A_{s,0}^L = 0.5$. If the $A_{s,0}$ of the VS model is different, then an adjusted throughput has to be used given by $(A_{s,0}/0.5) * X_{i,0}^L$. Therefore, the performance measurements of the adjusted FS neighbouring models are used according to the active secondary probabilities of (24-26).

To this end, the throughput, utilisation and mean queue length of the primary task are given by the following formulae:

$$X_{i,0}' = Z_L * d\left((A_{s,0}/0.5) * X_{i,0}^L\right) + Z_U * \left((A_{s,0}/0.5) * X_{i,0}^U\right), \tag{27}$$

$$U_{i,0}' = Z_L * \left((A_{s,0}/0.5) * U_{i,0}^L\right) + Z_U * \left((A_{s,0}/0.5) * U_{i,0}^U\right), \tag{28}$$

$$\langle n \rangle_{i,0}' = Z_L * \left((A_{s,0}/0.5) * \langle n \rangle_{i,0}^L\right) + Z_U * \left((A_{s,0}/0.5) * \langle n \rangle_{i,0}^U\right). \tag{29}$$

Developing the same pattern of analysis, the corresponding performance statistics for the secondary tasks of class $r = 1, 2, \ldots R_s^U$ can be estimated by the following expression

$$\begin{aligned} X_{i,r}' = \quad & S_L * \left(\left((A_{s,r} - A_{s,R_s^U})/0.5\right) X_{i,r}^L\right) + \\ & Z_U * \left(\left((A_{s,r} - A_{s,R_s^U} + (A_{s,R_s^U}/Z_U))/0.5\right) * X_{i,r}^U\right). \end{aligned} \tag{30}$$

Similar formulae can be devised to estimate the utilisation and mean queue length of the secondary tasks.

Station	Discipline	Class	0	1	2	3	4	5
1	PR	$\mu_{1,r}$	100	70	85	110	75	80
		$C_{s1,r}$	4	15	25	10	5	30
2	FCFS	$\mu_{2,r}$	30	15	25	40	35	20
		$C_{s2,r}$	3	7	4	2	8	10
3	LCFS	$\mu_{3,r}$	25	20	30	15	20	35
		$C_{s3,r}$	4	2	4	5	7	3
4	LCFS-PR	$\mu_{4,r}$	40	20	15	25	10	15
		$C_{s4,r}$	8	10	7	12	9	15
5	FCFS	$\mu_{5,r}$	15	40	20	30	15	10
		$C_{s5,r}$	5	2	1	7	10	4

Table 5.1 Raw data for QNM Configuration

Class r	$P_{r,1,2}$	$P_{r,1,3}$	$P_{r,1,4}$	$P_{r,1,5}$	$P_{r,syn}\{1\}$	$P_{r,syn}\{2\}$
0	0.30	0.20	0.15	0.35	0.90	0.50
1	0.50	0.30	0.10	0.10	0.90	0.50
2	0.20	0.20	0.30	0.30	0.90	0.50
3	0.10	0.40	0.40	0.10	0.90	0.50
4	0.20	0.30	0.10	0.40	0.90	0.50
5	0.15	0.25	0.25	0.35	0.90	0.50

Table 5.2 Routing and Synchronisation Probabilities

Model	$P_{r,syn}$	N	$V_{s,1}$	$V_{s,2}$	$V_{s,3}$	$V_{s,4}$	R'_s
M1	$\{1\}$	3	0.5	0.3	0.2	–	1.7
M2	$\{1\}$	3	0.3	0.4	0.2	0.1	2.1
M3	$\{1\}$	5	0.5	0.4	0.1	–	1.6
M4	$\{2\}$	4	0.4	0.3	0.3	–	1.9
M5	$\{2\}$	2	0.4	0.2	0.1	0.3	2.3
M6	$\{2\}$	3	0.5	0.2	0.3	–	1.8

Table 5.3 Secondary Splitting Probabilities per VS Model

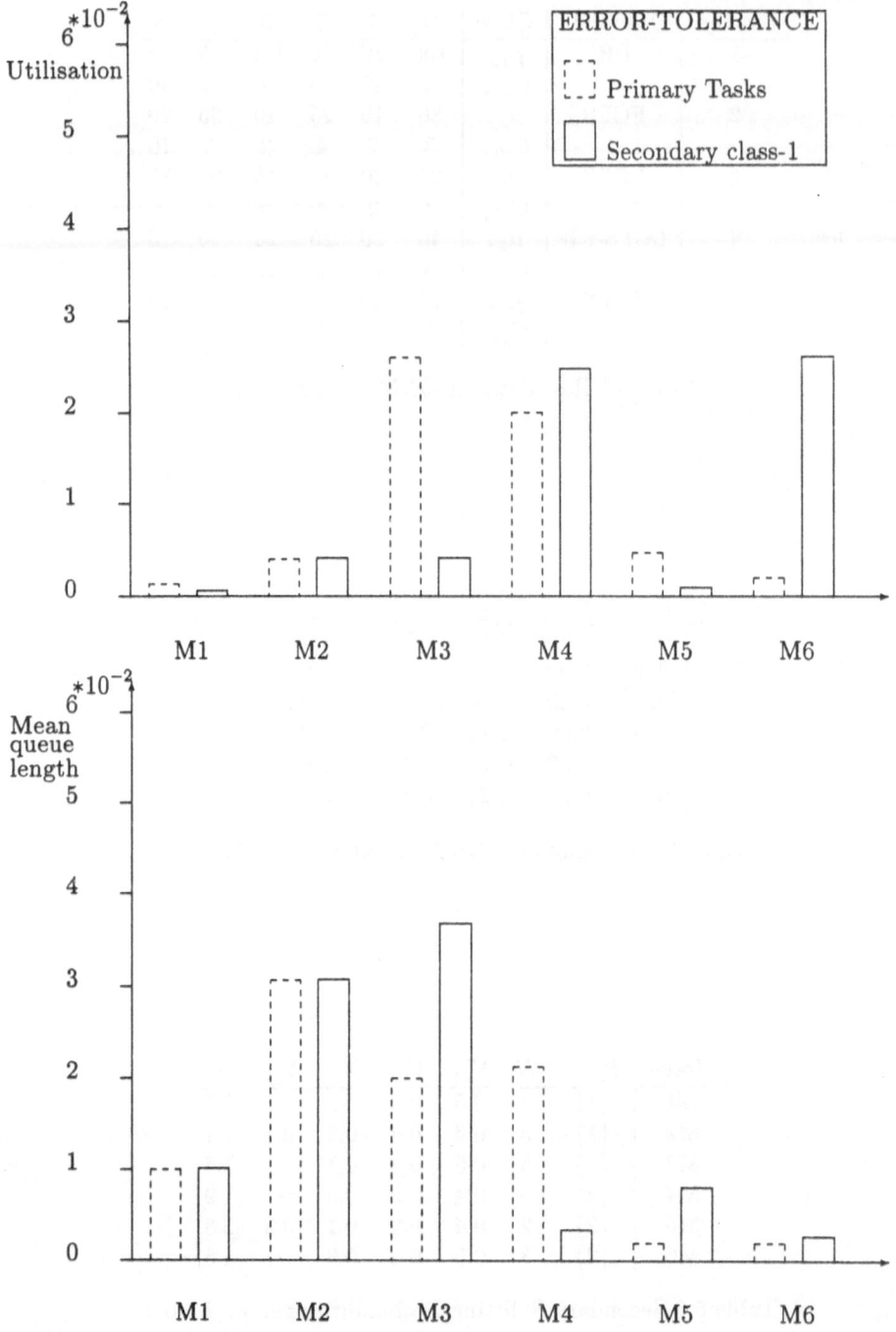

Figure 4: Utilisation and Mean-Queue-length Error-Tolerances of Primary Tasks Class 0 and Secondary Tasks Class 1 at CPU

5   Numerical validation

In this section a number of numerical experiments on the VS model are described to determine the relative accuracy of the ME-DELAY algorithm against simulation performed at the 95% confidence interval using QNAP-2 [20]. Although an explicit proof is lacking, several numerical studies have shown that the ME-DELAY algorithm always converges to a unique solution under various sets of input parameters and initial values.

The validation study of the VS model is focused, without loss of generality, on the central server QNM of Figure 2 with M=5. Tables 5.1 and 5.2 describe the parameters of the hardware configuration and the service time characteristics of the associated units, respectively. The same QNM is examined under infrequent (case {1}) or frequent (case {2}) synchronisation. Table 5.3 includes the secondary splitting probabilities of six VS models (M1-M6).

Figure 4 displays the error-tolerances for CPU utilisation and mean queue lengths associated with the primary and secondary class 1 tasks. It can be observed that the proposed approximation is very comparable in accuracy with that of simulation with error-tolerances less than 0.05. The computational cost for these experiments was on average one minute CPU time per model on the SUN4/280 computer. For more complex QNMs the maximum error-tolerance is always found to be less than 0.08, while their computational expense is dominated by that of the convolution-type procedure of the UME algorithm (step 3, Section 3.3) which is of

$$O\left(\left(M*N*(R_s+1)+2*(R_s+1)*(M-M^\dagger)\right)*N^{R_s+1)}\right),$$ where $M^\dagger$ is the

number of queueing stations under priority or IS service disciplines.

6   Conclusions and future work

In this paper the ME-DELAY approximate algorithm has been presented for the performance study of general QNMs with variable concurrency and synchronisation, generally distributed service times and mixed service disciplines. The credibility of the results is illustrated by carrying out relative comparisons against simulation. Future work includes the analysis of general QNMs with nested (multiple layers of) concurrent programming and variable concurrency in the layers.

References

[1] Ben-Ari M. Principles of concurrent and distributed programming. Prentice -Hall, Englewood Cliff, N.J., 1990.

[2] Whiddett D. Concurrent programming for software engineers. Ellis Horwood Ltd, Chichester, 1987.

[3] Peterson J. and Bulgren N. Studies in Markov models of computer systems. Proc. ACM Annual Conf., 1975, pp 102-107.

[4] Price T.G. Models of multiprogrammed computer systems with I/O buffering. Proc. 4th Texas Conf. Comp. Syst., Univ. Texas, 1975.

[5] Mackawa M. and Boyd D.L. Two models of task overlap within jobs of multi-processing multiprogramming systems. Proc. Int. Conf. Parallel Processing, 1976, pp 83-91.

[6] Towsley D.,Chandy K.M. and Browne J.C. Models for parallel processing with programs: Applications to CPU:I/O and I/O:I/O Overlap. Com. ACM, 1978, 21, pp 821-831.

[7] Herzog U., Hoffman W. and Kleinoder W. Performance Modelling and Evaluation for hierarchical organised multiprocessor computer systems. Proc. Int. Conf. Parallel Processing, 1979, pp 103-114.

[8] Bard Y. Some extension to multiclass queueing network analysis. Proc. 4th Int. Symp. Model. & Perfor. Eval. Comp. Syst., vol 1,1979.

[9] Sauer C.H. and Chandi K.M. Computer systems performance modelling. Prentice Hall, 1981.

[10] Heidelberger P. and Trivedi K.S. Analytic queueing models for programs with internal concurrency. IEEE Trans. Comp., 1983, C-32, pp 73-82.

[11] Thomaisan A. and Bay P.F. Analytic queueing network models for parallel processing of task systems. IEEE Trans. Comp., 1986, C-35, pp 1045-1054.

[12] Peng D. and Shin K.G. Modelling of concurrent task execution in distributed systems for real-time control. IEE Trans. Comp., 1987, C-36, pp 500-516.

[13] Skliros A.P. and Kouvatsos D.D. General queueing network models with job concurrency and synchronisation. Proc. 5th UK Comp and Telecom. Perf. Eng. Workshop, 1989, Univ. Edinburgh.

[14] Kouvatsos D.D. and Skliros A.P. General queueing network models of parallel task processing computer systems: A comparative study. Proc. 6th UK Comp and Telecom. Perf. Eng. Workshop, 1990, Univ. Bradford.

[15] Kouvatsos D.D. A universal maximum entropy algorithm for the analysis of the general closed networks. In: Hesagawa T. et al (eds) Computer Modelling and Performance Evaluation, North Holland, Amsterdam, 1986, pp 113-124.

[16] Jacobson P.A. and Lazowska E.D. Analyzing queueing networks with simultaneous resource possession. Com. ACM, 1981, pp 142-151

[17] U.S. Department of Defence. Programming language Ada: Reference manual, vol. 106, Lecture notes in Comp. Sci., Springer-Verlag 1981.

[18] Kouvatsos D.D. and Tabet-Aouel N.M. Product-form approximations for an extended class of general closed queueing networks. In: King P.J.B. et al (eds), Performance '90, North-Holland, Amsterdam, 1990, pp301-315.

[19] Georgatsos P.H. Modelling and analysis of computer communication networks with random or semidynamic routing, PhD Thesis, Univ. Bradford, 1989

[20] Veran M. and Potier D. A portable environment for queueing network modelling. In: Potier D. (ed), Modelling techniques and tools for performance analysis, North-Holland, Amsterdam, 1985, pp 25-63.

Experiences in Implementing Solution Techniques for Networks of Queues

David LL Thomas

British Telecom Computer
Performance Management Group
Cardiff

Abstract

This paper will show how to solve simple closed queueing network models using spreadsheets. Over the past decade there has been a growth in the use of personal computers and spreadsheets on the one hand and a growth in the use of approximate solution methods on the other. This paper will show that it is possible to use spreadsheets to re-create well known fixed point or approximate Mean Value Analysis results.

1 Introduction

The growth in use of spreadsheets has been phenomenal since they first burst onto the computing scene in the late 1970s. They have since become very sophisticated with their own macro language and graphics capabilities, in addition to having fast and efficient arithmetic. This paper gives the initial results of a short investigation into the uses of spreadsheets for solving closed queueing network modelling algorithms, using fixed point methods. The spreadsheet used was SuperCalc4 from Computer Associates. The PC used was an AT type without floating point chip.

We are attempting to see if the spreadsheet results are similar to the algorithms on which the equations are based. If we are able to do this we would be able to test new improved algorithms with a degree of confidence. We would also be able to build models quickly which incorporate features not found in commercial software (including our own). Our customers would have access to it in a form with which they are very experienced, and be able to incorporate the results into other packages.

2 Queueing Network Models

In the Computer Performance Management Group we act as computer performance specialists for the whole of British Telecom's commercial computing. To that end we are interested in solving models of computer systems for the tuning and capacity planning functions. We use analytical queueing network models for these purposes for speed and convenience.

We have built our own analytical modelling tool, called PANDORA. This runs on most IBM PCs and compatibles, and uses approximate Mean Value Analysis (MVA) algorithms. In building it we tested discrete parts of complex algorithms using spreadsheets. Before building PANDORA if our customers wanted to do their own 'what if' analysis we occasionally built simple single class open models of their systems, treating each device as an M/M/1 queue. This paper represents an extension of this type of work.

Large computer systems can be modelled accurately using an open network of queues, and a spreadsheet model of this can be done quite easily. However there are occasions where treating workloads as open can be inaccurate. Therefore we are also interested in being able to model closed networks of queues, using approximate methods.

3 Single Class Models

3.1 Open Response Time Equation

Building a model of a single workload is quite simple using the Response Time equation and Little's Law below:

$$R_k = S_k * (1 + ql_k) \qquad (1)$$

$$X = N/(R_{tot} + Z) \qquad (2)$$

$$ql_k = X * R_k \qquad (3)$$

where
R_k means response time at device k
R_{tot} means sum of device response times
S_k means service demand at device k
ql_k means queue length at device k
N means number of jobs or users in the workload
X means throughput and Z means think time.

The total response is calculated by summing the responses for each device. The think time is added to this to obtain the cycle time. The queue lengths for each device are obtained in equation 3 and are plugged straight back into equation 1. This is usually done by a recursive algorithm for exact MVA, and by iteration for approximate MVA. However the spreadsheet detects the 'circular' nature of the dependencies and converges to a solution. The spreadsheet layout is illustrated as example 1 in Appendix A. The number of terminals is 50 and the think time is 277.78 seconds. The service demand is 5 seconds at the CPU and 2.5 seconds at both of the disk drives.

The above model converges to the results below:

	Example 1		
	Spreadsheet	Exact	Error%
Response time	33.7	29.3	15
Throughput	0.161	0.163	-1
Q length cpu	4.07	3.42	19
Q length disk	0.670	0.674	-1

This example is taken from Lavenberg [1], and gives the same results as his algorithm. It also does for the rest of his single class results. The example above shows the worst results; most are markedly better.

3.2 Arrival Instant Equation

One of the problems with the response time equation as given in the first example is that the arriving transaction 'sees' itself in the queue. Lavenberg and Reiser [2] and Sevcik and Mitrani [3] addressed this problem with the arrival instant equation. In approximate MVA for single classes this takes the following form:

$$R_k = S_k * (1 + ql_k - (ql_k/N)) \tag{4}$$

In the spreadsheet we refer to ql_k/N as queue fraction qfk. The same results could have been obtained more simply by using $N - 1$ instead of N in equation 2.

The model converges to the results below:

	Example 2		
	Spreadsheet	Exact	Error%
Response time	32.1	29.3	10
Throughput	0.161	0.163	-1
Q length cpu	3.85	3.42	13
Q length disk	0.667	0.674	-1

This is an improvement on the previous method and the results, not surprisingly, are very similar to those of the Bard-Schweitzer Core (BSC) algorithm. The spreadsheet layout is illustrated as example 2 in Appendix A as Arrival Instant (AINST).

3.3 First Order Depth Improvement

One of the suggestions in Lavenberg's article was that it was worth investigating improving his approximations via a method similar to Lineariser, an algorithm developed by Chandy and Neuse [4]. This involves calculating the results of the model using BSC at the full population N and at the population of $N - 1$ (in this case 49). The differences between the queue lengths output for population $N - 1$ and those input to the response time equation at population N are plugged back into the model at population N. This is called First Order Depth Improvement (FODI) by van Doremalen et al [5].

The above model converges to the results below:

	Example 3		
	Spreadsheet	Exact	Error%
Response time	28.9	29.3	-2
Throughput	0.163	0.163	0
Q length cpu	3.36	3.42	-2
Q length disk	0.676	0.674	0

These results are very accurate, and are detailed in Appendix A, example 3 as FODI. They are similar to those of the Chandy-Neuse Lineariser (CNL) algorithm. In their paper Chandy and Neuse [4] run the Core algorithm at full population 3 times. Zahorjan et al [6] in their Aggregate Queue Length (AQL) algorithm run the Core algorithm at full population until successive runs converge.

Example 3 in Appendix A shows the calculation of the lineariser constant as in CNL, (which is more accurate than AQL). In this example the inner and outer convergence is done as in the AQL algorithm, in order to reduce the size of the model in the spreadsheet. The queue lengths are also aggregated as in AQL.

In his article Lavenberg [1] also gave examples of a single class model with 8 disk drives. The results for these also show the similarities between the above examples and the well known algorithms from the literature.

4 Multi-Class Models

4.1 Open Response Time Equation

The equations for these are as follows:

$$R_{kc} = S_{kc} * (1 + ql_k) \tag{5}$$

$$X_c = N_c/(R_{totc} + Z_c) \tag{6}$$

$$ql_{kc} = X_c * R_{kc} \tag{7}$$

where R_{kc} means response time at device k for workload c
R_{totc} means sum of device response times for workload c
S_{kc} means service demand at device k for workload c
ql_k means queue length at device k
ql_{kc} means queue length at device k for workload c
N_c means number of jobs or users in workload c
X_c means throughput and Z_c means think time for workload c

	Example 6		
	Workload 1		
	Spreadsheet	Exact	Error%
Response time	53.8	47.3	14
Throughput	0.0820	0.0829	-1
Q length cpu	2.29	1.86	23
Q length disk	1.07	1.03	4
	Workload 2		
	Spreadsheet	Exact	Error%
Response time	73.7	68.7	7
Throughput	0.0397	0.0400	-1
Q length cpu	1.94	1.75	11
Q length disk	0.492	0.499	-1

4.2 Improvements

This can be improved by changing the response time equation to the arrival instant version below without a perceptible slowing down of the solution.

$$R_{kc} = S_{kc} * (1 + q l_k - q f_{kc}) \qquad (8)$$

5 Preemptive Priorities

Some recent analyses of preemptive priority algorithms for closed networks start with Sevcik's Shadow Server (SDW) method. These include Agrawal [7], Bondi and Chuang [8], van Doremalen et al [5], Eager and Lipscomb [9] and Kaufman [10]. In this algorithm, only the workloads of the highest priority visit the CPU. All others visit 'Shadow CPUs', one for each priority level. The service times of the shadow CPUs are inflated by dividing by the proportion of time higher priority workloads are not visiting their respective servers.

5.1 Shadow CPU Method

We will take an example which comes from the book Metamodelling by Agrawal [7]. It is a simple model, but stresses preemptive priority algorithms. There are two workloads. The first, the high priority one, has four jobs. For the seven results shown in Appendix B the low priority workload has 1 to 7 jobs respectively. There are two devices. The first is a CPU which allows preemptive priority; the second, a disk drive which has a first in first out (FIFO) scheduling mechanism. The service demands are 3 for each workload device pair. Although the Shadow Server method was not designed with Mean Value Analysis in mind it suits it very well. The response time equations for the various device workload pairs are as follows:

$$R_{11} = S_{11}(1 + q l_{11} - q l_{11}/N_1) \qquad (9)$$

$$R_{12} = S_{12}(1 + q l_{12} - q l_{12}/N_2)/(1 - U_{11}) \qquad (10)$$

$$R_{21} = S_{21}(1 + q l_2 - q l_{21}/N_1) \qquad (11)$$

$$R_{22} = S_{22}(1 + q l_2 - q l_{22}/N_2) \qquad (12)$$

where U_{kc} is utilisation of device k by workload c.

This gives fair results for the high priority workload, but understates the response time at the CPU for the low priority workload. This in turn increases the low priority throughput. The results are shown in Appendix B as SDW.

5.2 The Completion Time Approximation (CTA) algorithm

The response time equation 10 (above) for the low priority workload at the CPU does not take preemption by the high priority jobs into account. This has been termed the delay error. van Doremalen et al [5] produced an algorithm which they called 'Completion Time Approximation' (CTA), which compensated for this. They tested CTA against algorithms which had also been tested in Metamodelling [7] by

Agrawal, and it seemed to be an improvement in many cases. In this example this has the effect of adding $S11 * ql11$ to equation 10 as in equation 13 below.

$$R_{12} = S_{12}(1 + ql_{12} - ql_{12}/N_2)/(1 - U_{11}) + (S_{11} * ql_{11}) \tag{13}$$

The general form of this equation is given in van Doremalen et al [5] and Bondi and Chuang [8]. This gives the improved figures shown in Appendix B as CTA.

5.3   Correction of the Synchronisation Error

There are still some structural errors not solved by CTA. The main one is called the synchronisation error. This is due to the failure of some algorithms to take into account the fact that when a low priority job is at the CPU, then all higher priority jobs must be elsewhere; in this case at the disk drive.

This is taken into account in this particular example by either of the following equivalent equations:

$$R_{22} = S_{22}(1 + ql_2 + ql_{11} - qf_{22}) \tag{14}$$

$$R_{22} = S_{22}(1 + ql_{22} + N_1 - qf_{22}) \tag{15}$$

These both add the queue length at the CPU for the high priority workload to the disk drive for the low priority workload. Bondi and Chuang [8] say that this can overestimate the number of jobs seen by an arriving job at that server. This is accompanied by an increase in the interarrival time variability of low priority jobs at the disk drive. Their suggestion to counteract this would lead to the following amendment to equation 11 in this particular example:

$$R_{21} = S_{21}(1 + ql_2 - qf_{22} - qf_{21}) \tag{16}$$

These amendments lead to the results marked SYNCH in Appendix B. All of these results could expect to be improved on using a FODI method. We also include the results of an algorithm by Kouvatsos and Tabet-Aouel [11] which are the most consistently accurate we have encountered.

6   Non MVA Based Algorithms

6.1   Single Class Models

If the response time equation for open workloads in example 1 is replaced by the equivalent equation 17, and equations 2 and 3 by 18, the same results are obtained, but more quickly. However, at high utilisations there are convergence problems.

$$R_k = 1/((1/S_k) - X) \tag{17}$$

$$X = N/(R_{tot} + Z) \tag{18}$$

There are fewer dependencies in the above equations, and the number of equations per iteration is in the order K where K is the number of devices. The above equations are similar to those in Kelly [12]. The single MVA algorithm is in the order $2K$ equations per iteration.

The spreadsheet version is shown as example 4 in Appendix A. It produces the same results as example 1 for the models in Appendix A.

6.2   Multi-Class Models

We do not use throughput in the response time equation, but use utilisation, which is throughput multiplied by service demands.

$$R_{kc} = S_{kc}/(1 - U_k) \tag{19}$$

$$X_c = N_c/(R_{tot} + Z_c) \tag{20}$$

$$U_{kc} = X_c * S_{kc} \tag{21}$$

Although there are in the order of $2K$ equations per iteration, there are fewer dependencies than equations 5 to 7. There can be convergence problems at high utilisations, and sometimes on convergence the results are infeasible. The results are the same as using equations 5 to 7, but are obtained more quickly.

6.3   Accuracy

The results of approximations for large models are of interest, as these cannot be solved by exact methods. The amount of time and space required by exact methods increases with the number of devices, workloads and the number of jobs per workload.

The approximate methods used do not increase their time and space requirements with the number of jobs per workload. The ones using the response time equation for open workloads were in part designed to be more accurate when there were large numbers of jobs per workload.

6.4   Possible Improvements

It is worth while investigating whether we can improve on the accuracy of the above equations in a way analogous to Bard-Schweitzer. For single workload models this can be done simply by changing equation 18 to:

$$X = (N - 1)/(R_{tot} + Z) \tag{22}$$

This produces the same results as example 2 for single workloads, and is slightly more robust. Like BSC it has the property of producing exact results in a single class model when the MPL is 1. It also seems to have the property, with BSC, of producing exact results when the models have balanced workloads i.e. when the work for each workload is balanced equally over the devices.

| | Example 2 | | |
	Spreadsheet	Exact	Error%
Response time	32.1	29.3	10
Throughput	0.161	0.163	-1
Q length cpu	3.85	3.42	13
Q length disk	0.667	0.674	-1

7 Summary

7.1 Results

Using variations of the Response Time equation and Little's Law we can use fixed point methods to solve large single and multi-class models quickly and easily using spreadsheets. More modern hardware and software would speed up the results by at least one order of magnitude. The results are very similar to algorithms developed for the purpose of solving these models.

By substituting the arrival instant equation to calculate response times, the results are more accurate, especially with workloads with a small number of jobs. These results are very similar to those of the Bard-Schweitzer Core algorithm.

When using First Order Depth Improvement methods, the results are very accurate, and similar to those of the Chandy-Neuse Lineariser for the examples calculated.

7.2 Problems

There are two main difficulties in using spreadsheets in developing algorithms or models. They are tracing of bugs and convergence problems.

Error tracing In developing algorithms for use in computer software we have been able to use traces, which can be browsed or printed, at various points within the software to highlight or track down errors. This facility does not seem to be available with the spreadsheets we have been using. Single stepping through an iteration seems to be available, but has not yet been used.

Convergence Algorithms which depend on dividing by $(1 - U)$ or service rate less throughput, can lead to failure to converge, or results outside the feasible region (such as negative response times), especially if the utilisation is high.

7.3 Further Work

We have only used the basic facilities of spreadsheets. New versions contain the abilities to do 3D graphics and have powerful scripting languages. The ability to create multi-dimensional worksheets may aid solution by 'decomposition' methods. We will endeavour to increase the number and types of model and algorithm we can build using spreadsheets. Examples could include:

1. modelling of memory constraints by Multi-Programming levels (MPLs)

2. extension of FODI method to preemptive priority algorithms

3. using the spreadsheet to apply numerical methods to the solutions.

7.4 Conclusions

It is possible to use variations on approximate Mean Value Analysis (AMVA) equations to obtain fast, accurate results for some queueing network models using spreadsheets on PCs. The models using AMVA equations are robust, converge quickly and the methods used are simple.

Spreadsheets enable the analyst to build and test algorithms and models without recourse to programming languages. This goes some way to reducing the 'semantic gap' between equations, algorithms and programming code, and could prove useful in speeding up development of new algorithms and models.

Although queueing network modelling software is being used by hundreds of performance analysts in the U.K., the cheapest software is expensive and considered difficult to understand, if not use. Spreadsheet packages are orders of magnitude cheaper and used by millions of people. The equations used in MVA are simple algebraically and conceptually. They can be made visible in spreadsheets, and the underlying dependencies uncovered by movement of a cursor or mouse.

The use of spreadsheets for queueing network modelling therefore may be of use to the whole computing community; the developers of modelling software, the performance analysts, and their customers to their mutual benefit.

References

[1] Lavenberg S.S. Closed multichain product form queueing networks with large population sizes. In: Disney R L, Ott T J, (eds) Applied Probability. Vol.1. Birkhauser; 1981, pp 219-249.

[2] Lavenberg S. S. and Reiser M. Stationery state probabilities at arrival instants for closed queueing networks with multiple types of customers. Journal of Applied Probability. Vol. 17; 1980, pp 1048-1061.

[3] Sevcik K. C. and Mitrani I. The distribution of queueing network states at input and output instants. Journal of the ACM. 28; 1981, pp 358-371.

[4] Chandy K. M. and Neuse D. Fast accurate heuristic algorithms for queueing network models of computing systems. Communications of the ACM 25,2; Feb. 1982, pp 126-134.

[5] van Doremalen J., Wessels J. and Wijbrands R. Approximate Analysis of Priority Queueing Networks. Teletraffic Analysis and Computer Performance Evaluation. Science; 1986, pp 117-131.

[6] Zahorjan J., Eager D. L. and Sweillam H. M. Accuracy, Speed, and Convergence of Approximate Mean Value Analysis. Performance Evaluation 8; 1988, pp 255-270.

[7] Agrawal S. C. Metamodelling. MIT Press, 1984.

[8] Bondi A. B. and Chuang Y-M. A New MVA-Based Approximation for Queueing Networks with a Preemptive Priority Server. Performance Evaluation 8; 1988, pp 195-221.

[9] Eager D. L. and Lipscomb J. N. The AMVA Priority Approximation. Performance Evaluation 8; 1988, pp 173-193.

[10] Kaufman J. S. Approximation methods for networks of queues with priorities. Performance Evaluation 4; 1984, pp 183-198.

82

[11] Kouvatsos D. and Tabet-Aouel N. Personal communication. 1989.

[12] Kelly F. P. On a Class of Approximations For Closed Queuing Networks. Queuing Systems: Theory and Applications. 1988

Appendix A

Cell	Contents	Cell	Contents
(a1)	No. of terminals	(b1)	50
(a2)	Think time	(b2)	277.78
(a3)	CPU service demand	(b3)	5
(a4)	Disk1 service demand	(b4)	2.5
(a5)	Disk2 service demand	(b5)	2.5
(a6)		(b6)	
(a7)	Response cpu	(b7)	sum(b3*(1+b15))
(b8)	Response disk1	(b8)	sum(b4*(1+b16))
(a9)	Response disk2	(b9)	sum(b5*(1+b17))
(a10)	tot resp	(b10)	sum(b7.b9)
(a11)	cycle	(b11)	sum(b10+b2)
(a12)		(b12)	
(a13)	xput	(b13)	sum(b1/b11)
(a14)		(b14)	
(a15)	q length cpu	(b15)	sum(b13*b7)
(a16)	q length disk1	(b16)	sum(b13*b8)
(a17)	q length disk2	(b17)	sum(b13*b9)

Example 1 - Using Open Response time = S(1 + ql) equation

Cell	Contents	Cell	Contents
(a1)	No. of terminals	(b1)	50
(a2)	Think time	(b2)	277.78
(a3)	CPU service demand	(b3)	5
(a4)	Disk1 service demand	(b4)	2.5
(a5)	Disk2 service demand	(b5)	2.5
(a6)		(b6)	
(a7)	Response cpu	(b7)	sum(b3*(1+b15-b19))
(b8)	Response disk1	(b8)	sum(b4*(1+b16-b20))
(a9)	Response disk2	(b9)	sum(b5*(1+b17-b21))
(a10)	tot resp	(b10)	sum(b7.b9)
(a11)	cycle	(b11)	sum(b10+b2)
(a12)		(b12)	
(a13)	xput	(b13)	sum(b1/b11)
(a14)		(b14)	
(a15)	q length cpu	(b15)	sum(b13*b7)
(a16)	q length disk1	(b16)	sum(b13*b8)
(a17)	q length disk2	(b17)	sum(b13*b9)
(a18)		(b18)	
(a19)	q fraction cpu	(b19)	sum(b15/b1)
(a20)	q fraction disk1	(b20)	sum(b16/b1)
(a21)	q fraction disk2	(b21)	sum(b17/b1)

Example 2 - Using Arrival Instant Equation

Cell	Contents	Cell	Contents
(a1)	No. of terminals	(b1)	50
(a2)	Think time	(b2)	277.78
(a3)	CPU service demand	(b3)	5
(a4)	Disk1 service demand	(b4)	2.5
(a5)	Disk2 service demand	(b5)	2.5
(a6)		(b6)	
(a7)	Response cpu	(b7)	sum(b3*(1+b15-b19+b27))
(b8)	Response disk1	(b8)	sum(b4*(1+b16-b20+b28))
(a9)	Response disk2	(b9)	sum(b5*(1+b17-b21+b29))
(a10)	tot resp	(b10)	sum(b7.b9)
(a11)	cycle	(b11)	sum(b10+b2)
(a12)		(b12)	
(a13)	xput	(b13)	sum(b1/b11)
(a14)		(b14)	
(a15)	q length cpu	(b15)	sum(b13*b7)
(a16)	q length disk1	(b16)	sum(b13*b8)
(a17)	q length disk2	(b17)	sum(b13*b9)
(a18)		(b18)	
(a19)	q fraction cpu	(b19)	sum(b15/b1)
(a20)	q fraction disk1	(b20)	sum(b16/b1)
(a21)	q fraction disk2	(b21)	sum(b17/b1)
(a22)		(b22)	
(a23)	diff ql cpu	(b23)	sum(b45-b15)
(a24)	diff ql disk1	(b24)	sum(b46-b16)
(a25)	diff ql cpu	(b25)	sum(b47-b17)
(a26)		(b26)	
(a27)	lin const cpu	(b27)	sum(b1-1)*b23
(a28)	lin const disk1	(b28)	sum(b1-1)*b24
(a29)	lin const disk2	(b29)	sum(b1-1)*b25
(a30)		(b30)	
(a31)	No. of terminals	(b31)	49
(a32)	Think time	(b32)	277.78
(a33)	CPU service demand	(b33)	5
(a34)	Disc1 service demand	(b34)	2.5
(a35)	Disc2 service demand	(b35)	2.5
(a36)		(b36)	
(a37)	Response cpu	(b37)	sum(b33*(1+b45-b49+b53))
(b38)	Response disk1	(b38)	sum(b34*(1+b46-b50+b54))
(a39)	Response disk2	(b39)	sum(b35*(1+b47-b51+b55))
(a40)	tot resp	(b40)	sum(b37.b39)
(a41)	cycle	(b41)	sum(b40+b32)
(a42)		(b42)	
(a43)	xput	(b43)	sum(b31/b41)
(a44)		(b44)	

Example 3 - Using Arrival Instant Equation and FODI (first part)

Cell	Contents	Cell	Contents
(a45)	q length cpu	(b45)	sum(b43*b37)
(a46)	q length disk1	(b46)	sum(b43*b38)
(a47)	q length disk2	(b47)	sum(b43*b39)
(a48)		(b48)	
(a49)	q fraction cpu	(b49)	sum(b45/b31)
(a50)	q fraction disk1	(b50)	sum(b46/b31)
(a51)	q fraction disk2	(b51)	sum(b47/b31)
(a52)		(b52)	
(a53)	lin const cpu	(b53)	sum(b1-2)*b23
(a54)	lin const disk1	(b54)	sum(b1-2)*b24
(a55)	lin const disk2	(b55)	sum(b1-2)*b25

Example 3 - Using Arrival Instant Equation and FODI (continued)

Cell	Contents	Cell	Contents
(a1)	No. of terminals	(b1)	50
(a2)	Think time	(b2)	277.78
(a3)	CPU service rate	(b3)	1/5
(a4)	Disk1 service rate	(b4)	1/2.5
(a5)	Disk2 service rate	(b5)	1/2.5
(a6)		(b6)	
(a7)	Response cpu	(b7)	sum(1/(b3-b13))
(b8)	Response disk1	(b8)	sum(1/(b4-b13))
(a9)	Response disk2	(b9)	sum(1/(b5-b13))
(a10)	tot resp	(b10)	sum(b7.b9)
(a11)	cycle	(b11)	sum(b10+b2)
(a12)		(b12)	
(a13)	xput	(b13)	sum(b1/b11)
(a14)		(b14)	

Example 4 - Using Response $= 1/((1/S)-X)$ equation

Cell	Contents	Cell	Contents
(a1)	No. of terminals	(b1)	50
(a2)	Think time	(b2)	277.78
(a3)	CPU service rate	(b3)	1/5
(a4)	Disk1 service rate	(b4)	1/2.5
(a5)	Disk2 service rate	(b5)	1/2.5
(a6)		(b6)	
(a7)	Response cpu	(b7)	sum(1/(b3-b13))
(b8)	Response disk1	(b8)	sum(1/(b4-b13))
(a9)	Response disk2	(b9)	sum(1/(b5-b13))
(a10)	tot resp	(b10)	sum(b7.b9)
(a11)	cycle	(b11)	sum(b10+b2)
(a12)		(b12)	
(a13)	xput	(b13)	sum((b1-1)/b11)
(a14)		(b14)	

Example 5 - Using $X = N - 1/R_{tot}$

Cell	Contents	Cell	Contents
(a1)	wkl hi no. of jobs	(b1)	4
(a2)	wkl lo no. of jobs	(b2)	4
(a3)	wkl hi cpu service	(b3)	3
(a4)	wkl hi disk service	(b4)	3
(a5)	wkl lo cpu service	(b5)	3
(a6)	wkl lo disk service	(b6)	3
(a7)		(b7)	
(b8)	wkl hi resp cpu	(b8)	sum(b3*(1+b20-b26))
(a9)	wkl hi resp disk	(b9)	sum(b4*(1+b24-b27))
(a10)	wkl hi resp tot	(b10)	sum(b8.b9)
(a11)	wkl lo resp cpu	(b11	sum(b5*(1+b22-b28)/(1-b18))
(a12)	wkl lo resp disk	(b12)	sum(b6*(1+b24-b29))
(a13)	wkl lo resp tot	(b13)	sum(b11.b12)
(a14)		(b14)	
(a15)	wkl hi xput	(b15)	sum(b1/b10)
(a16)	wkl lo xput	(b16)	sum(b2/b13)
(a17)		(b17)	
(a18)	wkl hi cpu util	(b18)	sum(b15*b3)
(a19)	.	(b19)	
(a20)	wkl hi ql cpu	(b20)	sum(b15*b8)
(a21)	wkl hi ql disk	(b21)	sum(b15*b9)
(a22)	wkl lo ql cpu	(b22)	sum(b16*b11)
(a23)	wkl lo ql disk	(b23)	sum(b16*b12)
(a24)	tot ql disk	(b24)	sum(b21+b23)
(a25)		(b25)	
(a26)	wkl hi qf cpu	(b26)	sum(b20/b1)
(a27)	wkl hi qf disk	(b27)	sum(b21/b1)
(a28)	wkl lo qf cpu	(b28)	sum(b22/b2)
(a29)	wkl lo qf disk	(b29)	sum(b23/b2)

Example 6 - Using Shadow CPU method

as shadow CPU example above except as follows:

Cell	Contents	Cell	Contents
(a11)	wkl lo resp cpu	(b11)	sum((b5*(1+b22-b28)/(1-b18))+b3*b20)

Example 7 - Using CTA approximation

as CTA example above except as follows:

Cell	Contents	Cell	Contents
(a9)	wkl hi resp disk	(b9)	sum(b4*(1+b24-b29-b27))
(a12)	wkl lo resp disk	(b12)	sum(b6*(1+b24+b20-b29))

Example 8 - Correction of synchronisation error

Appendix B - Comparison of Different Models for Preemptive Priority Systems

Model6 has 2 workloads; high priority (H) and low (L). The two devices are a CPU and disk drive. Workload H has an MPL of 4 while L has an MPL of from 1 to 7. The service times are 3 seconds for each workload/device pair.

| | Class H Throughput Comparison | | | | | | | | |
| | Throughput | | | | | % Error | | | |
NL	GB	SDW	CTA	SYNCH	KT-A	SDW	CTA	SYNCH	KT-A
1	0.2552	0.2434	0.2482	0.2667	0.2604	-4.6	-2.7	4.5	2.0
2	0.2438	0.2248	0.2331	0.2464	0.2502	-7.8	-4.4	1.1	2.6
3	0.2330	0.2092	0.2200	0.2292	0.2398	-10.2	-5.6	-1.6	2.9
4	0.2229	0.1955	0.2083	0.2141	0.2300	-12.3	-6.6	-3.9	3.2
5	0.2136	0.1835	0.1975	0.2008	0.2213	-14.1	-7.5	-6.0	3.6
6	0.2050	0.1728	0.1876	0.1889	0.2137	-15.7	-8.5	-7.9	4.2
7	0.1970	0.1632	0.1784	0.1782	0.2071	-17.2	-9.4	-9.5	5.1

| | Class L Throughput Comparison | | | | | | | | |
| | Throughput | | | | | % Error | | | |
NL	GB	SDW	CTA	SYNCH	KT-A	SDW	CTA	SYNCH	KT-A
1	0.0226	0.0470	0.0374	0.0278	0.0228	108.0	65.5	23.0	0.9
2	0.0419	0.0756	0.0616	0.0506	0.0436	80.4	47.0	20.8	4.1
3	0.0587	0.0969	0.0803	0.0703	0.0621	65.1	36.8	19.8	5.8
4	0.0734	0.1140	0.0959	0.0876	0.0781	55.3	30.7	19.3	6.4
5	0.0864	0.1285	0.1094	0.1030	0.0917	48.7	26.6	19.2	6.1
6	0.0981	0.1411	0.1215	0.1167	0.1031	43.8	23.9	19.0	5.1
7	0.1085	0.1521	0.1324	0.1291	0.1126	40.2	22.0	19.0	3.3

GB	Global Balance (Exact) Solution, from QNAP2 Harmondsworth.
SDW	Sevcik's Shadow CPU algorithm
CTA	Completion Time Approximation
SYNCH	CTA with amendment to cope with synchronisation problem
KT-A	Kouvatsos / Tabet-Aouel, personal communication

The global balance solution was calculated by the QNAP package, assuming the disk drive to be a FIFO server. The results in Agrawal [7] assume a Processor Sharing (PS) discipline at the non preemptive server.

BCMP Queueing Networks versus Stochastic Petri Nets: A Pragmatic Comparison

Peter J.B. King and Saqer Abdel-Rahim
Department of Computer Science
Heriot-Watt University

Abstract

Queueing network solution packages that analyse the performance of BCMP networks are widely available, and are one of the standard tools of the performance analyst. Stochastic Petri net analysis packages are less commonly used. They have only recently emerged from the research laboratory, and are not in widespread commercial use. This paper takes a particular problem, diskless workstation configuration management, and analyses it using tools of both types. Although the theoretical limitations of both types of solution package are well known, little work has been done to compare their practical performance.

1 Introduction

Since Buzen's publication of a fast algorithm for the normalisation constant of an exponential queueing network[3] and the seminal BCMP paper[1], product form queueing networks have been the tool of first resort for most performance analysts. With suitable parameterisation, they can provide an accurate and speedy analysis of a performance model. A steadily increasing number of commercial packages implementing product form models have been produced[11, 9, 2]. The different packages usually provide the same basic solution techniques, but differ in the amount of pre and post analysis of results that they provide. Some will take manufacturer's accounting data and analyse it to produce the input parameters to a model directly. Others provide the heuristic extensions that allow priority queues and non-exponential servers to be analysed. A number of new algorithms have been discovered, but none have significantly improved on the basic complexity of Buzen's convolution algorithm.

In the mid 1980s packages based on the concept of a stochastic Petri net became available. Petri nets were originally developed to analyse the correctness of parallel algorithms, and took a time independent view of the world: the choice of which event of those that were able to occur actually took place was taken arbitrarily, and the probability of its occurrence had no effect on the analysis. This is of course the appropriate model when correctness is the concern. By attaching rates and probabilities to the events in a system, some estimate of the performance can be made. If all events either occur instantaneously, or take a time which is exponentially distributed, then a stochastic Petri net is equivalent to a Markov process and can be transformed into one. In general, stochastic Petri nets offer no theoretical advantages over Markov processes, but they provide a standard notation, understood by other engineers for expressing performance models. The availability of standard Markov process analysis tools allows the calculation of both steady state and transient solutions for the Petri net. Sensitivity analysis can also be performed.

The limitations of the theory underpinning both Petri nets and BCMP queueing networks are well understood, and their respective limitations do not form the subject of our investigation. We concentrate on the practical consequences of the theoretical limits, investigating the changes in the solution times as successively refined models are solved by both techniques.

First we describe the example problem that we are going to analyse. The theoretical limitations of BCMP queueing models are then outlined and a simple model developed in our chosen BCMP solution package. The same process is then followed for a stochastic Petri net model.

2   Diskless Workstation Configuration

The advent of powerful desktop workstations has shifted the problem of hardware configuration from being "how fast a machine should I buy" to "where should I place resources". The system designer has to decide on where to place file servers in the local area network, whether to have single or multiple servers, whether to split the network into a number of segments linked by bridges, etc.

The sample problem that we investigate has workstations which are performing various similar file updating tasks on a common set of files. We assume that there are two groups of workstations, which have different access patterns to the files. For the purposes of this model we assume that the files also fall into two groups. This allows us to investigate the effects of splitting the files between two disks on the same server, and using two file servers. As a further refinement we investigate the effect of separating the Ethernet into two halves, with a bridge server to join them, and allocating each group of workstations its own Ethernet. One file server would be placed on each Ethernet, containing the most popular file group for that group of workstations. One sample configuration is shown in Figure 1, with A and B representing workstations of the appropriate class, and F representing the fileserver.

3   Queueing Network Model

Networks of queues with servers which give exponentially distributed service times were known to have a particularly simple form of solution, known as product form. This was found by Jackson[7] and rediscovered independently by Gordon and Newell for closed networks, that is networks with a fixed number of customers[6]. Their result was little used, however, because of the computational cost of calculating the normalisation constant. Buzen[3] published a fast algorithm for calculating the normalisation constant, and showed how other performance metrics could be derived simply once the normalisation constant was known. The paper of Baskett, Chandy, Muntz, and Palacios-Gomez[1] has since been further generalised, but it remains the theoretical basis on which most queueing network packages are founded. They state the restrictions on networks, job types, and routing, necessary to have a network

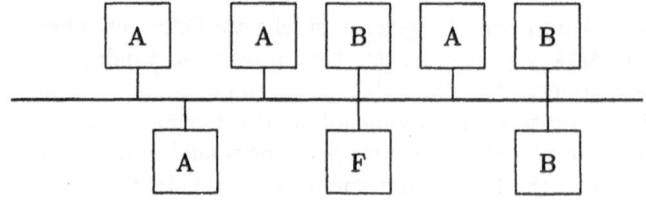

Figure 1: Diskless Workstation System

with a product form solution. Incorporating those later extensions to the admissible models that we use, the restrictions on the networks can be summarised. Jobs which circulate around the network can be of different classes. Each class of job can have its own routing probabilities between servers in the network, and jobs can also change class. The routing and class change behaviour of jobs is accommodated by a set of probabilities. A job of class r, on leaving node i in the network, will go to node j and be of class s with probability $p_{ir,sj}$. It is possible for some classes to be *open*, that is allow external arrivals and departures, whilst others are *closed*, with a fixed number of jobs circulating.

The server nodes in the network can be of a number of different types too. Processor sharing servers give all jobs simultaneous service, dividing their service equally amongst all the jobs present at the server. This is the continuous analogue of a round robin CPU scheduler, with the CPU quantum taken infinitesimally small. At last come first served, preemptive resume, servers a newly arriving job displaces a job that is currently in service and starts execution immediately. If it is itself interrupted by subsequent arrivals, it will resume at the point of interruption when all subsequently arriving jobs have been processed. First come first servers are the traditional queue. Random order of service queues take any of their waiting jobs for service. Infinite server, otherwise known as delay servers, effectively give each job its own server. For FCFS servers and random order of service servers, each job must have the same exponential service demand, independent of class. At servers of other types, the service demands of jobs can depend on the class of the job. In all sorts of server the total rate of service can depend on the number of jobs queued at the server.

It may not surprise the reader to learn that after the discovery of Buzen's convolution algorithm, a number of improvements and also some radically different algorithms for solving the same class of networks have been developed. They are of roughly equivalent theoretical complexity, although some are conceptually easier to explain, and one has a significantly better complexity when the number of classes of jobs is large. A number of heuristic techniques for selecting the order of incorporation of servers into the solution can be used to improve the performance of other techniques. Most commercial packages implement a large subset at least of the types of server and network that are admissible for product form solution using

BCMP theorems and their extensions. Some commercial packages incorporate the heuristic approximate extensions for priority networks, and other server types that are not admissible under BCMP.

The package that we have used for the purposes of this paper is Panacea[9], which was developed at Bell Laboratories. As well as the standard convolution algorithm it implements an approximation technique based on asymptotic expansions of multiple integrals which has excellent properties for large population networks. Panacea supports most of the standard types of server in a BCMP network, but the version which we have used does not support variable rate servers where the service rate depends on queue length. Multiple job classes are supported, and jobs can switch class. For queueing networks, the Ethernet poses a problem, since traffic passing both to and from the file server must pass through the Ethernet, and will interfere with traffic in both directions. A simple queueing network, without any job classes cannot describe this phenomenon, since, the routing must be only dependent on the server just visited. If job classes are available, then different job classes can be used to distinguish between traffic in the two different directions at the Ethernet. The Ethernet server in the queueing network is represented by a processor sharing node[1] Ideally, the service rate would be queue length dependent, but our version of Panacea does not support this facility. We incorporated a loop in the Panacea program to estimate the mean queue length and use the transmission rate that was implied by that mean queue length.

Job classes are also used to distinguish between jobs originating at the different workstations, since they have different file access behaviour and must return to the correct workstation after they have been served by the fileserver.

A representation of the model is shown in Figure 2.

4   SPNP Model

Petri nets were originally a graphical representation for concurrent processes, used to describe their synchronisation and deadlock properties. Extensive research has been done on these aspects of Petri nets, and various extensions have also been proposed. The first use of Petri nets to evaluate performance appears to have been reported by Sifakis[10], and the thesis of Molloy[8] was another milestone. Since then, research on stochastic Petri nets has blossomed.

A Petri net may be defined as a graph with two types of nodes, *places*, and *transitions*. They are linked by directed arcs, with the arcs joining a places to transition, or transitions to places, but *not* places to places or transitions to transitions. Places can be *marked*, and transitions can *fire* removing a token from each input place, and adding a token to each output place. The original Petri net used this time independent description. Stochastic Petri nets use the same definitions, except that transitions only fire after a random delay once they are enabled. Usually this delay is exponentially distributed. It is obvious that such stochastic Petri nets are equivalent to Markov processes, and provide a standardised way of generating such models for solution.

---

[1]In this case, since all requests to the server are identical in length, a processor sharing server will give the same probabilistic behaviour as a random order of service discipline.

92

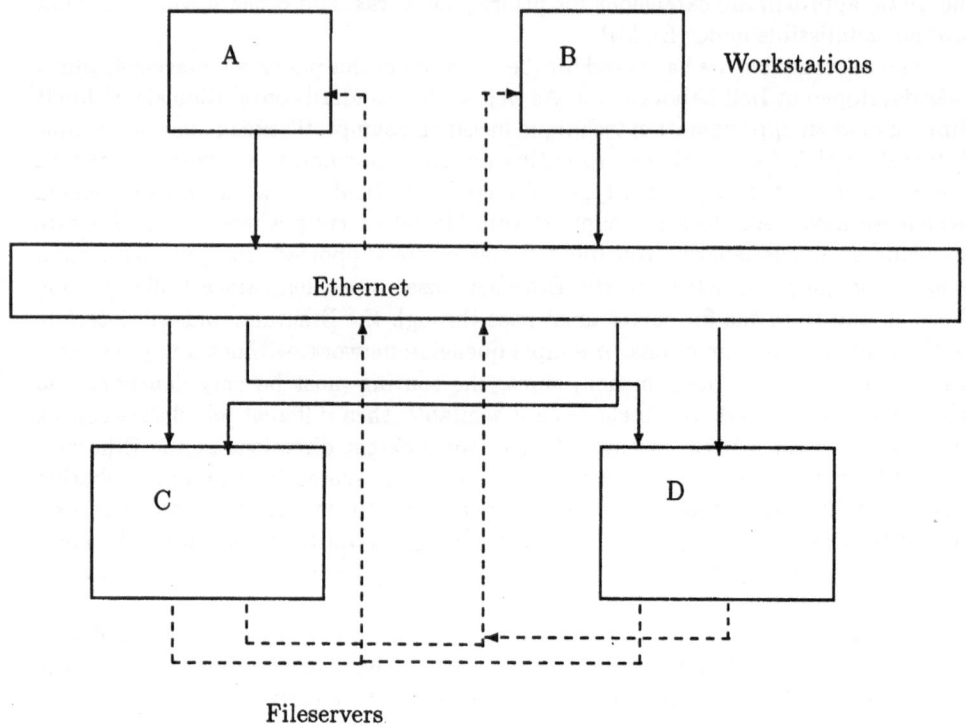

Figure 2: Queueing Network Model

The package which we use to specify and solve stochastic Petri net models is SPNP[4]. This package is implemented as a set of C functions, which the user augments with C functions of his own that define the Petri net. It is possible to define arrays of places and transitions, and of the arcs between them. The arcs can be standard arcs, or have variable multiplicity, that is, remove a different number of tokens from the input place, depending on the marking. Inhibitor arcs, which prevent transitions from firing when the place to which they are linked is marked also exist. The transitions in SPNP nets can either be instantaneous, or timed.

Our diskless workstation configuration can be modelled in a number of seemingly equivalent fashions, but they have dramatically different complexities when it comes to solving them.

It is simplest to explain the initial model in terms of a single class of workstations and a single fileserver. The initial model that was developed is shown in Figure 3. The large circles represent places in the Petri net. Simple open boxes represent timed transitions, and the boxes with a horizontal line represent instantaneous transitions. Arcs are represented by arrows, with inhibitor arcs being represented by the lines from places to transitions ending in a small circle. A token represents each workstation. The places are used to record the state of the workstations. When the tokens leave the thinking place, W, they contend for the Ethernet and then are served by a file server, place F, before returning to the thinking place, using the Ethernet again. The Ethernet also needs places representing the tokens which are in transit from file server to client as well as from client to file server. Attempting to model the Ethernet as closely as possible, the initial model distinguishs the workstation which has actually captured the Ether from those which are contending for it. An immediate transition separates the place representing the contending stations from the one representing the workstation actually in possession of the Ether. These transitions were protected by inhibitor arcs to prevent more than one token being in the Ethernet possession places at any time. The probability of the instantaneous transition firing is proportional to the numbers of jobs contending for the Ether. Since the job which succeeds in acquiring the Ether is essentially chosen at random from those contending, the probability of the instantaneous transition firing is the ratio of the number of tokens in the input place to the transition to the total number of tokens in the E.w1 and E.w2 places. Tokens leaving the workstations place, W, go to place E.w1, where they remain as long as E.g1 or E.g2 is occupied. There is at most one token in E.g1 and E.g2. There cannot be a token in both simultaneously. When transition EtoF or EtoW fires, if there are any tokens in E.w1 or E.w2, one of the instantaneous transitions, I.1, or I.2, fires, and one token is moved to the E.g1 or E.g2 place. The inhibitor arcs prevent the instantaneous transitions form firing again until the E.g1 or E.g2 place is empty again. The rate at which the transition removing tokens from the Ethernet fires is related to the number of jobs contending for the Ethernet. This is to account for the acquisition delay caused by the initial contention for the Ether. The transition rate is calculated by adding the length of the transmission, after the contention period is finished, to the contention period. The length of the contention period can be found as follows. Assume that the probability of a station which is contending for the Ether transmitting in any slot is p, then the probability that one and only one of k stations contending for the Ether will transmit is given by $q = p(1-p)^{k-1}$. The number of slots in a contention period has a modified geometric distribution with parameter $(1-q)$, and has mean

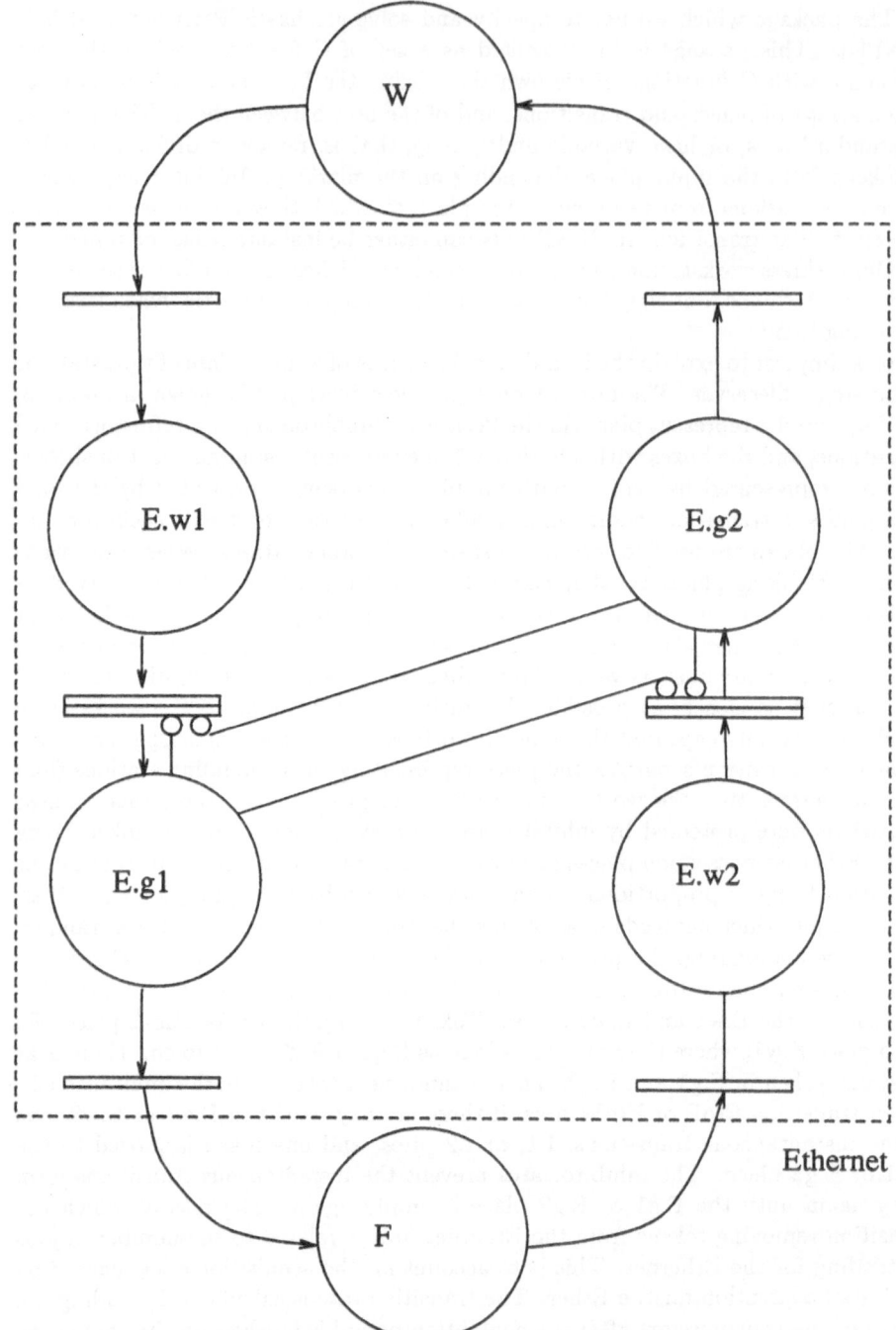

Figure 3: Stochastic Petri Net Model

$1/q$. However the last slot in the contention period is part of the transmission, so $(1-q)/q$ slots are added to the transmission period to find the mean time for which a station will occupy the Ether, and hence its mean transmission rate. The probability of a station transmitting in a particular slot during a contention period depends on the randomisation algorithm used by the Ethernet protocol. The binary exponential backoff algorithm gives a probability of 2^{-k+1}. We have actually used a probability of $1/k$ in our experiments, which is an over estimate for large k, but with small populations will give almost identical results.

When different populations of jobs are involved, and they are destined to different workstations, an array of places is used , one place representing each combination of workstation/fileserver. Thus when there are two of each, the Ethernet is represented by 16 places, with 8 immediate internal transitions, and 64 inhibitor arcs to enforce the restriction that only one station is transmitting at any time.

This technique for modelling the Ethernet accurately represents what happens, but unfortunately it also generates a very large reachability graph, and an equally large number of states for the Markov chain. The combinatorial explosion in the state space as the number of workstations increases is such that it can only be solved for a system with two workstations of each class! By combining the waiting and actually transmitting places in the Ethernet, eliminating the instantaneous transitions, and scaling the firing rate of the transitions that leave the Ethernet, a very similar model can be built. Now the Ethernet has only 8 places, and firing rate for leaving these places is the same as it was in the previous model, scaled by i/m when there are i jobs in the place being left, and m jobs at the Ethernet in total. This modification enables the SPNP model to be solved for a system with up to 8 workstations in total. These are the results presented in the following section. In the current model, where the lengths of file server requests are independent of the class of workstation, and of the destination of the request, it is highly likely that the second model is merely a lumped version of the first one, although we have not attempted to prove this. Further reductions in the state space can be made by recognising that because the request length is independent of the destination or source of the request, there is no need to commit to a particular destination until after the Ethernet has been traversed. The Ethernet can then be represented by 4 places, corresponding to requests from the two classes of workstation, inbound and outbound from the fileservers. The places representing the Ethernet inbound to the fileservers then have two output transitions, which fire at the rate of the old transitions off the Ethernet, except that they are multiplied by the probability of choosing the corresponding fileserver. This allows the model to be served when ther are 12 workstations in total.

5 Results

The basic model that we study has two file servers and two groups of workstations. The results from the Panacea model are given in Table 1. Each workstation has a mean think time of 0.2 seconds (this is really the time between successive file requests). Group 1 of workstations access group 1 files with probability 0.8, and group 2 files otherwise. Group 2 workstations access group 1 files with probability 0.3 and group 2 files with probability 0.7. The workstations are evenly distributed

96

between the two groups. It can be seen that as the load increases, the response time for workstations of group 1 increases faster than that for group 2. This reflects the groups more intensive use of the file server with the group 1 files, which is the bottleneck device in the system. The SPNP model gives identical results, until it runs out of space when 10 workstations are used. This is hardly surprising since both models are built on completely Markovian assumptions. The CPU time consumed is of some interest. It is known that the convolution method for solving queueing network models is basically linear in the number of workstations. Using Panacea, there was no increase in execution time for increasing number of workstations. Clearly this implies that for smallish models the overhead of parsing and setting up the description of the model overwhelms the solution time. With SPNP, the times were as follows: 2 workstations, 0.1 sec; 4 workstations, 1.9 sec; 6 workstations, 16 sec; 8 workstations 127 sec. The number of states in the Markov chain were 49, 784, 7056, and 44100, respectively. Unfortunately, it is not possible to distinguish between the phases of constructing the transition matrix of the Markov chain and solving the resulting steady state equations, but for the example tried, the former appears to dominate[2].

The other two tables give the results for a single file server with two disks attached, and for a system with the file servers on separate Ethernets with a bridge between them. It can be seen that the bridge reduces the contention for each Ethernet considerably, but since it was not the bottleneck in the first place, only tiny improvements in response time for high loads, and deterioration in response time at low loads. These models have not been run with SPNP yet, but no significant differences would be expected.

This comparison of queueing networks and stochastic Petri nets has been rather one sided. The advantages of a BCMP solver are overwhelming when it can be used. Here it can. A Petri net solution could be more accurate in situations where the BCMP model had to be an approximation, eg if the file requests to the different file servers were of different sizes. There are a number of classes of Petri net that admit specila forms of solution. For example, Florin and Natkin[5] have found a class which exhibit "vector product form". It seems likely that our model falls in this class, but without a specialised solver to take advantage of the structure, the possible advantages of this structure cannot be converted into increased solution speed.

References

[1] F. Baskett, K.M. Chandy, R.R. Muntz, and F. Palacios-Gomez. Open, closed and mixed networks of queues with different classes of customers. *Journal of the ACM*, 22(2):248–260, 1975.

[2] M. Booyens, P.S. Kritzinger, A.E. Krzesinski, P. Teunissen, and S. van Wyk. SNAP: An analytical multiclass queueing network analyser. In D. Potier, editor, *Modelling Techniques and Tools for Performance Analysis*, pages 67–79. North Holland, Amsterdam, The Netherlands, 1985.

[2]Messages are issued by SPNP at the start and end of each phase, but there is no timing information given

Table 1: Two File Servers

WS's	Utilisation			Throughput			Response	
	server1	server2	Ether	server1	server2	Ether	ws1	ws2
2	0.13	0.10	0.06	4.6	3.8	17.0	0.035	0.035
4	0.25	0.20	0.11	9.2	7.5	33.5	0.039	0.038
6	0.36	0.30	0.17	13.5	11.1	49.2	0.044	0.043
8	0.47	0.39	0.22	17.6	14.5	64.2	0.050	0.048
10	0.57	0.47	0.27	21.4	17.6	78.1	0.057	0.055
12	0.67	0.55	0.31	24.9	21.5	90.7	0.067	0.062
14	0.75	0.62	0.35	27.9	23.1	101.9	0.078	0.072
16	0.81	0.68	0.38	30.4	25.3	111.5	0.092	0.082
18	0.87	0.73	0.41	32.5	27.2	119.4	0.108	0.095
20	0.91	0.77	0.43	34.1	28.7	125.6	0.128	0.109

Table 2: One File Server with Two Disks

WS's	Utilisation				Throughput				Response	
	serv	disk1	disk2	Ether	serv	disk1	disk2	Ether	ws1	ws2
2	0.13	0.12	0.10	0.05	15.9	4.4	3.5	15.9	0.052	0.052
4	0.25	0.23	0.19	0.10	31.0	8.5	7.0	31.0	0.058	0.058
6	0.36	0.33	0.27	0.15	45.3	12.4	10.2	45.3	0.065	0.064
8	0.47	0.43	0.35	0.20	58.5	16.1	13.2	58.5	0.074	0.073
10	0.56	0.52	0.43	0.24	70.5	19.4	15.9	70.5	0.085	0.083
12	0.65	0.60	0.49	0.28	81.2	22.3	18.3	81.2	0.097	0.094
14	0.72	0.66	0.55	0.31	90.3	24.6	20.4	90.3	0.112	0.107
16	0.78	0.71	0.59	0.34	98.0	26.8	22.2	98.0	0.129	0.123
18	0.83	0.76	0.63	0.36	104.3	28.5	23.6	104.3	0.150	0.140
20	0.87	0.83	0.66	0.37	109.3	29.8	24.8	109.3	0.172	0.160

Table 3: Two Ethernets, Bridged

WS's	Utilisation				Throughput				Response	
	serv1	serv2	Eth1	Eth2	serv1	serv2	Eth1	Eth2	ws1	ws2
2	0.12	0.10	0.04	0.03	4.7	3.8	11.0	10.1	0.037	0.037
4	0.24	0.20	0.07	0.07	9.2	7.5	21.6	19.9	0.040	0.040
6	0.36	0.30	0.11	0.10	13.5	11.0	31.8	29.4	0.045	0.045
8	0.47	0.38	0.14	0.13	17.5	14.4	41.5	38.4	0.050	0.050
10	0.57	0.47	0.17	0.16	21.4	17.5	50.5	46.8	0.058	0.056
12	0.66	0.54	0.20	0.19	24.8	20.4	58.6	54.5	0.066	0.063
14	0.75	0.61	0.23	0.21	27.8	23.0	65.8	61.5	0.078	0.072
16	0.81	0.67	0.25	0.23	30.4	25.3	71.8	67.5	0.092	0.083
18	0.87	0.73	0.26	0.25	32.5	27.2	76.7	72.6	0.108	0.095
20	0.91	0.77	0.28	0.26	34.1	28.7	80.4	76.9	0.128	0.109

[3] J.P. Buzen. Computational algorithms for closed queueing networks with exponential servers. *Communications of the ACM*, 16(9):527–531, September 1973.

[4] G. Ciardo, J.K. Muppala, and K.S. Trivedi. SPNP: Stochastic Petri net package. In *Proceedings of 3rd International Workshop on Petri Nets and Performance Models*, pages 142–151, Kyoto, Japan, 1989.

[5] G. Florin and S. Natkin. Generalization of queueing network product form solutions to stochastic petri nets. *IEEE Transactions on Software Engineering*, SE-17(2):99–107, February 1991.

[6] W.J. Gordon and G.F. Newell. Closed queueing networks with exponential servers. *Oper. Res.*, 15(2):254–265, 1967.

[7] J.R. Jackson. Jobshop-like queueing systems. *Manage. Sci.*, 10(1):131–142, 1963.

[8] M.K. Molloy. Performance analysis using stochastic Petri nets. *IEEE Transactions on Computers*, C-31(9):913–917, 1982.

[9] K.G. Ramakrishnan and D. Mitra. An overview of PANACEA, a software package for analyzing Markovian queueing networks. *BSTJ*, 61(10):2849–2872, 1982.

[10] J. Sifakis. Use of Petri nets for performance evaluation. In H. Beilner and E. Gelenbe, editors, *Measuring, Modelling and Evaluating Computer Systems*, pages 75–93. North-Holland, Amsterdam, The Netherlands, 1977.

[11] M. Veran and D. Potier. QNAP 2: A portable environment for queueing systems modelling. In D. Potier, editor, *Modelling Techniques and Tools for Performance Analysis*, pages 25–63. North Holland, Amsterdam, The Netherlands, 1985.

Extension to Preemption of a Method for a Feedback Queue

D.H.J. Epema
Department of Mathematics and Computer Science
Delft University of Technology
P.O. Box 356, 2600 AJ Delft
The Netherlands

Abstract

We consider a time-sharing model with a single server, multiple queues, priorities, and feedback to lower priority queues, which, disregarding the service discipline, is of type M/G/1. In the model, there is a finite or countably infinite number K of queues. Jobs arrive at queue 1 according to a Poisson process. A fraction q_{k+1} of jobs is appended to queue $k+1$ after having received service in queue k; the remainder leaves the system. Feedback may or may not depend on service time. Jobs in queue k have priority over jobs in queue l, when $k < l$. In [7], Schrage introduces a method to derive the Laplace-Stieltjes transform of the total residence time in queues $1, 2, \ldots, j$ of jobs that ever reach queue j, and Wolff [9] uses the same method to derive the total delay in queues $1, 2, \ldots, j$ in the special case of one overall service time and quantum allocation, both in the nonpreemptive case. In this paper, this method is extended to the preemptive discipline, and analogous results are obtained.

1 Introduction

In [9], Wolff develops a very elegant method to derive the mean waiting times in an M/G/1 queueing model with multiple queues, priorities, and feedback. In this model, there is a finite or countably infinite number K of queues. On arrival, jobs join the tail of queue 1 with some service demand. Jobs are served in passes, receiving a quantum δ_k on their pass through queue k. If this satisfies their remaining service demand, they leave the system, otherwise they are appended to queue $k+1$. Jobs in queue k have priority over jobs in queue l, when $k < l$. Scheduling is nonpreemptive. The average total delay d_j of jobs in their first j passes (i.e., the sum of the waiting times in queues $1, 2, \ldots, j$) is obtained by considering a modified model in which the first j queues are replaced by one queue with quantum $\sum_{i=1} j\delta_i$, deriving the average delay d_J in this queue, and subsequently correcting d_J for busy periods induced by arrivals occurring during d_J and the total service $\sum_{i=1} j - 1\delta_i$ received by these jobs on their first $j - 1$ passes in the original model.

In Schrage [7], the same method is used to obtain the Laplace-Stieltjes transform (LST) of the probability distribution function (pdf) of the total residence time in

100

queues $1, 2, \ldots, j$. In fact, a slightly more general model is considered there: in every queue, jobs have a separate, general service demand, and a fraction q_{k+1} of jobs in queue k are appended to queue $k+1$, the remainder leaving the system.

In this paper we extend this method to the preemptive discipline, which is definitely more useful in time-sharing computer systems, and obtain the LST of the pdf of the total residence time in queues $1, 2, \ldots, j$ in this general model, and by differentiation an expression for its mean. As a corollary, we obtain in the special case of quantum allocation a formula for the average total delay experienced by jobs in queues $1, 2, \ldots, j$, dependent on the amount of service required in queue j; the average total delay of jobs needing exactly j passes; and the average total delay in queues $1, 2, \ldots, j$ of jobs needing at least j passes.

This extension to preemption can, in a somewhat different setting and not in the full generality as presented here, also be found in Kleinrock [5], p. 174-5. There, a foreground-background (FB) scheduling algorithm with quantum size tending to zero is considered: when the system is non-empty, all jobs with equal smallest received service time share the server. This model is preemptive in that a newly arriving job gets the server all to himself until he catches up with the job(s) receiving service on his arrival. With the same method as in [7] and in the present paper, the LST of the total residence time conditioned on service time is obtained. Also, as the FCFS special case of the general model in [5], Section 4.7, the mean waiting time in the ordinary preemptive quantum allocation case (with finite quanta) conditioned on service time is derived along the same lines (ibid., p. 180-1).

For another interesting paper on quantum allocation algorithms, in which in particular nonpreemptive feedback to lower priority queues is treated in a way similar to [7], see Brown [1]. Results on mean residence and waiting times for the model of this paper (and related or much more general models) can be found in [3, 6, 8].

2 The Queueing Model

We consider the M/G/1 queueing model defined thus:

1. There are K queues, numbered $1, 2, \ldots$, with K finite or countably infinite, and there is one server. Jobs arrive at queue 1 according to a Poisson process with rate λ.

2. In queue k, jobs have a generally and independently distributed service time S_k with pdf $P(S_k \leq t) = G_k(t)$, and mean s_k.

3. A fraction q_{k+1} of the jobs in queue k join queue $k+1$ with a new service demand drawn from S_{k+1} after having completed their k-th service; the others leave the system, $k = 1, 2, \ldots, K$. We put $q_1 = 1$. The probability of a particular job in queue k of joining queue $k+1$ is either completely random (Bernoulli feedback), or only depends on random properties of the job (e.g., its service demand in queue k; see the discussion at the end of this section). Let S_k', S_k'' be the service time in queue k of jobs that proceed to queue $k+1$ or leave the system from queue k, with pdfs $G_k'(t), G_k''(t)$ and means s_k', s_k'', respectively. Obviously, $G_k(t) = q_{k+1} G_k'(t) + (1 - q_{k+1}) G_k''(t)$.

4. Jobs in queue k have priority over jobs in queue l, when $k < l$. Scheduling is preemptive resume, referred to in this paper simply as preemptive.

Throughout this paper this model is denoted by M.

Let $p_k = q_k \cdot q_{k-1} \cdot \ldots \cdot q_1$ be the fraction of *all* jobs that ever reach queue k, and let $\lambda_k = p_k \lambda$ be the arrival rate in queue k, $k = 1, 2, \ldots, K$. In order for the system to be stable, we require $\sum \lambda_k s_k < 1$.

In the definition of the model, we require the feedback to depend on random properties of jobs (in fact, in our model service time in a particular queue is the only such property). The case where feedback is completely random is generally referred to in the literature as Bernoulli feedback (BF). Then, $S'_k = S''_k = S_k$, $k = 1, 2, \ldots, K$.

As another special case we will consider (preemptive) Quantum Allocation (QA). Then, jobs have a generally and independently distributed service time S when they initially arrive at queue 1. Let $P(S \le t) = G(t)$. In queue k, a job receives a quantum of service of size δ_k and is fed back to the tail of queue $k + 1$, or it receives its remaining service demand if this is less than or equal to δ_k, and leaves the system. This fits the general model by putting:

$$G_k(t) = \frac{G(t + \Delta_{k-1}) - G(\Delta_{k-1})}{G\,c(\Delta_{k-1})}, \quad 0 \le t < \delta_k,$$

$$G_k(t) = 1, \qquad\qquad t \ge \delta_k,$$

where $\Delta_k = \sum_{i=1} k\delta_i$, $\Delta_0 = 0$, and $G\,c(t) = 1 - G(t)$ for $t \ge 0$. We require $\sum \delta_k = \infty$. In particular, if $K < \infty$, we assume $\delta_k < \infty, k = 1, 2, \ldots, K - 1$, and $\delta_K = \infty$. In QA-case, $p_k = G\,c(\Delta_{k-1})$, $S'_k \equiv \delta_k$, and the pdf $G''_k(t)$ of S''_k is

$$G''_k(t) = \frac{G(t + \Delta_{k-1}) - G(\Delta_{k-1})}{G(\Delta_k) - G(\Delta_{k-1})}, \quad 0 \le t < \delta_k,$$

$$G''_k(t) = 1, \qquad\qquad t \ge \delta_k.$$

In the definition of queueing models, often all quantities are assumed to be independent. This is not always necessary. In our model, the feedback decision may depend on a random property of a job, which in this case can only be its service time. With BF, when it is a job's turn to be served, random, independent draws from the appropriate distributions determine its service time and whether it is fed back. In the QA-case, we can imagine that when a job starts receiving its first service in queue k, first a random decision for feedback is made; if the job is fed back (probability q_{k+1}), it gets a complete quantum δ_k, and otherwise a random draw of S''_k. This can easily be incorporated in the analysis, as is shown in Section 4. Of course, in the model with QA, the situation at feedback instants is 'worse' than in the model with BF, if the same S_k and q_k (derived from an overall service time) are used, because in QA the longest jobs continue.

In the nonpreemptive case, the analysis for time-dependent feedback can be carried a step further, for then the state $(n_1, n_2, \ldots, n_K)$ of the system in terms of the numbers n_i of jobs in queue $i, i = 1, 2, \ldots, K$ at departure instants (defined as the instants when a single service in some queue and the entailing feedback have completed) forms an embedded Markov chain. This can be seen as follows. Suppose that for a job requiring an amount t of service in queue $k-1$, the feedback probability of going from queue $k-1$ to queue k is $q_k(t)$, $k = 2, 3, \ldots, K$, with $\int_0^\infty q_k(t)\, dt = q_k$. Then the transition probability from state $(0, \ldots, 0, n_k, n_{k+1}, \ldots, n_K)$ to state $(n, 0, \ldots, 0, n_k - 1, n_{k+1} + 1, \ldots, n_K)$, where $n_k > 0$, is

$$\int_0^\infty e^{-\lambda t} \frac{(\lambda t)^n}{n!} q_{k+1}(t)\, dG_k(t);$$

similarly, the transition probability from state $(0, \ldots, 0, n_k, n_{k+1}, \ldots, n_K)$ to state $(n, 0, \ldots, 0, n_k - 1, n_{k+1}, \ldots, n_K)$ is

$$\int_0^\infty e^{-\lambda t} \frac{(\lambda t)^n}{n!} (1 - q_{k+1}(t))\, dG_k(t).$$

In the QA-case, $q_{k+1}(t) = 0$ for $0 \le t < \delta_k$, and $q_{k+1}(t) = 1$ for $t \ge \delta_k$, so then these probabilities are

$$\int_{\delta_k -}^{\delta_k +} e^{-\lambda t} \frac{(\lambda t)^n}{n!}\, dG_k(t) = e^{-\lambda \delta_k} \frac{(\lambda \delta_k)^n}{n!} q_{k+1},$$

and $e^{-\lambda \delta_k} ((\lambda \delta_k)^n / n!)(1 - q_{k+1})$, respectively, because $G_k(t)$ has an impulse of area q_{k+1} at $t = \delta_k$.

In the BF-case, we simply have $q_k(t) \equiv q_k$.

In the preemptive case, in general the state of the system at departure instants in terms of the numbers of jobs is not a Markov chain. When at a departure instant the state is $(0, \ldots, 0, n_k, \ldots, n_K)$, with $n_k > 0$, the first job to depart may either be the one at the head of queue k, say job A, or, if A is preempted, a job in queue 1. However, this preemption depends on the remaining service of A, and this in turn depends on the number of times A has been preempted. So except when all $G_k(\cdot)$ are negative exponential, we do not have a Markov chain.

3 Preliminaries

We introduce the following notation and terminology:

- A job in queue k is called a k-pass. It can be in three states: waiting before it has received any service, receiving service, or interrupted.

- A job which leaves the system after completing its service in queue k is called a k-job.

- The total service demand in queues $1, 2, \ldots, k$ of *all* jobs is denoted by $\overline{S}_k$; let its pdf be $\overline{G}_k(\cdot)$ and its mean $\overline{s}_k$. It depends on the feedback regime how $\overline{S}_k$ and $\overline{G}_k(\cdot)$ are composed of $S_1, S_2, \ldots, S_k$ and $G_1, G_2, \ldots, G_k$, respectively. In the BF-case,

$$\overline{G}_k(t) = \sum_{i=1}^{k-1} p_i (1 - q_{i+1})(G_1(t) * G_2(t) * \ldots * G_i(t)) + p_k(G_1(t) * G_2(t) * \ldots * G_k(t)), \quad k \ge 1,$$

where $G * H$ is the convolution product of G and H.

In the QA-case, $\overline{S}_k = \min(S, \Delta_k)$, and $\overline{G}_k(\cdot)$ is $G(\cdot)$ truncated at Δ_k:

$$\overline{G}_k(t) = G(t), \quad 0 \le t < \Delta_k,$$
$$\overline{G}_k(t) = 1, \quad t \ge \Delta_k.$$

Of course, $\overline{G}_k(\cdot)$ can also be written as a composition of the functions $G_k(\cdot)$ defined in section 2 for the QA-case. Note that this is quite another composition than in the BF-case. In particular, in the QA-case, $\overline{G}_k(\cdot)$ has a longer tail (but the same mean) than in the BF-case, if the same $G_k(\cdot)$ were used.

- Let $\overline{B}_k(\cdot)$ be the pdf of the busy period duration of the first come first served (FCFS) M/G/1 queue with service demand $\overline{S}_k$.

- Let $\overline{S'_k}$ be the total service time in queues $1, 2, \ldots, k$ of jobs that reach at least queue $k + 1$, with pdf $\overline{G'_k}(\cdot)$ and mean $\overline{s'_k}$, that is, $\overline{S'_k} = S'_1 + S'_2 + \ldots + S'_k$. In the QA-case, $\overline{S'_k} \equiv \Delta_k$.

- The LST of a pdf $H(\cdot)$ is denoted by $H * (s) = \int_0^\infty e^{-st} dH(t)$. If $H(\cdot)$ is the pdf of the stochastic variable T, then its mean is $E[T] = -dH * /ds(0)$.

- In the FCFS M/G/1 queue with arrival rate λ, service time S with pdf $G(\cdot)$, and traffic intensity $\rho = \lambda E[S]$, the LST of the waiting time distribution $W(\cdot)$ is given by the Pollaczek-Khintchine (P-K) transform equation (cf. [4], p. 200):

$$W * (s) = \frac{s(1-\rho)}{s - \lambda + \lambda G * (s)} \tag{1}$$

Differentiation yields the P-K mean value formula for the mean waiting time w:

$$w = \frac{\lambda E[S^2]}{2(1-\rho)} \tag{2}$$

- Let $H(\cdot)$ be the pdf of some real, non-negative stochastic variable T (e.g., some waiting or service time), and let T_I be T interrupted by busy periods of the FCFS M/G/1 queue with arrival rate λ, service time S with pdf $G(\cdot)$, service rate μ, traffic intensity $\rho = \lambda/\mu$, and busy period duration pdf $B(\cdot)$. Following the reasoning as in [2], pp. 175-6, the LST of the pdf $H_I(\cdot)$ of T_I is given by

$$H_I * (s) = H * (s + \lambda - \lambda B * (s)) \tag{3}$$

In particular, the LST of $B(\cdot)$ satisfies the busy period relation (ibid.):

$$B * (s) = G * (s + \lambda - \lambda B * (s)) \tag{4}$$

Differentiation of (4) and (3) yields $\beta = 1/(\mu(1-\rho))$ for the mean busy period length β and $E[T_I] = E[T]/(1-\rho)$, respectively. That is, such interruptions induce an expansion by a period of length

$$E[T_I] - E[T] = \frac{\rho E[T]}{1-\rho}. \tag{5}$$

- The total traffic intensity due to jobs in queues $1, 2, \ldots, k$ is $\rho_k = \sum_{i=1}^{k} k\lambda_i s_i$, $k = 1, 2, \ldots, K$. Put $\rho_0 = 0$.

- w_k is the mean waiting time of k-passes, which is defined as the time elapsed between the moment the k-pass joins queue k, and the moment it receives its first service.

- The total delay of a job is defined as its total residence time in the system minus its total service demand.

4 Analysis

The main goal of this section is to prove the following theorem (put $\overline{B}_0 * (\cdot), \overline{G}'_0 * (\cdot) \equiv 1$).

Theorem. (a) *The probability distribution function $R_j(t)$ of the total residence time in queues $1, 2, \ldots, j$ of jobs that reach queue j in M has LST*

$$
\begin{aligned}
R_j * (s) \;=\; & W_J * (s + \lambda - \lambda \overline{B}_{j-1} * (s)) \cdot \overline{G}'_{j-1} * (s + \lambda - \lambda \overline{B}_{j-1} * (s)) \cdot \\
& G_j * (s + \lambda - \lambda \overline{B}_{j-1} * (s)),
\end{aligned}
\tag{6}
$$

with

$$
W *_J (s) = \frac{s(1 - \rho_j)}{s - \lambda + \lambda \overline{G}_j * (s)}.
\tag{7}
$$

(b) *The mean total residence time r_j in queues $1, 2, \ldots, j$ of jobs that reach queue j in M is*

$$
r_j = \frac{\lambda E[\overline{S}_j\, 2] + 2(1 - \rho_j)(\overline{s}'_{j-1} + s_j)}{2(1 - \rho_{j-1})(1 - \rho_j)}
\tag{8}
$$

Proof. (a) Following the reasoning in [9] and [7], consider the modified M_j, in which queues $1, 2, \ldots, j$ are replaced by one queue denoted by J, for any fixed j, $1 \leq j \leq K$. Jobs arrive at queue J, have service demand equal to their total service demand in queues $1, 2, \ldots, j$ in the original model M, are served in FCFS fashion (uninterrupted), and the same jobs as in M (fraction p_{j+1}) subsequently join queue $j + 1$. For any realization and at any epoch, the server is either working on the same job in the same queue k, $k > j$, in either model, or on some k-pass with $k \leq j$ in M and on a J-pass in M_j.

Because of preemption, queue J is simply a FCFS M/G/1 queue with arrival rate λ and service demand $\overline{S}_j$, so by (1) the LST of the waiting time distribution $W_J(\cdot)$ in queue J is given by (7).

The LST of the total residence time of all jobs in queue J is $W_J * (s) \cdot \overline{G}_j * (s)$, and the LST of the total residence time in queue J of jobs that would reach queue j in M, say R_J, is $W_J * (s) \cdot \overline{G}'_{j-1} * (s) \cdot G_j * (s)$.

In order to find the total residence time in queues $1, 2, \ldots, j$ in M of jobs that reach queue j, R_J has to be corrected for interruptions by busy periods due to arrivals occurring during R_J (that is, including service in queue j). These are FCFS M/G/1 busy periods with arrival rate λ and service time pdf $\overline{G}_{j-1}(\cdot)$, so application of (3) yields (6).

(b) Denoting by w_J the mean waiting time in queue J in M_j, differentiation of (6) and (7) gives

$$w_J = \frac{\lambda E[\overline{S}_j\ 2]}{2(1 - \rho_j)} \tag{9}$$

and

$$r_j = \frac{w_J + \overline{s}'_{j-1} + s_j}{1 - \rho_{j-1}} \tag{10}$$

These together prove (b).

Remark. (a) Returning to the point of the feedback mechanism, even though formulas (6,7,8) are the same in every case, they depend on this mechanism because of $\overline{G}_j(\cdot)$, $\overline{G}'_{j-1}(\cdot)$, $E[\overline{S}_j\ 2]$, and $\overline{s}'_{j-1}$. However, as long as the same $G_k(\cdot)$ and q_k are used, the busy periods in queues $1, 2, \ldots, j - 1$ with pdf $\overline{B}_{j-1}(\cdot)$ are the same, regardless of which jobs are fed back.

(b) The LST of the total residence time in the system of j-jobs and its mean is obtained by replacing $G_j * (\cdot)$ by $G''_j * (\cdot)$ in (6), and s_j by s''_j in (8).

(c) Equations (6,7) should be compared to Eq. (4.28) in [5], which represents the analogous result for the FB-discipline as explained in the introduction. Eq. (8) in case of QA is equivalent to [5], Eq. (4.35).

Corollary. *In the case of quantum allocation, the expected total delay $d_j(\delta)$ in M of a j-job requiring δ service in queue j, $0 < \delta \leq \delta_j$, is*

$$d_j(\delta) = \frac{\lambda E[\overline{S}_j\ 2] + 2\rho_{j-1}(1 - \rho_j)(\Delta_{j-1} + \delta)}{2(1 - \rho_{j-1})(1 - \rho_j)}, \qquad j = 1, 2, \ldots \tag{11}$$

Proof. Apply the theorem (b) with $r_j = d_j(\delta) + \overline{s}'_{j-1} + s_j$, $\overline{s}'_{j-1} = \Delta_{j-1}$, and $s_j = \delta$.

As special cases of (11), $d_j(0)$ is the average total delay of jobs that reach queue j until they receive their first service in queue j, $d_j(s_j)$ is the average total delay of such jobs in queues $1, 2, \ldots, j$, and $d_j(s''_j)$ is the average total delay of j-jobs.

In [3], a very general queueing model with multiple queues *and* job classes, priorities, and feedback is studied, of which the QA-model is a special case. There, we obtain the following recursive expression for the mean waiting times:

$$w_j = \frac{\lambda G\ c(\Delta_{j-1}) E[S_j\ 2]}{2(1 - \rho_{j-1})(1 - \rho_j)} + \frac{\lambda \int_{\Delta_{j-2}} \Delta_j G\ c(t) dt \cdot \Sigma_{i=1}\ j - 1 t_i}{1 - \rho_j}, \qquad j = 1, 2, \ldots \tag{12}$$

Here, $t_i = w_i + \delta_i/(1 - \rho_{i-1})$ is the mean total time in queue i of jobs with service demand at least Δ_i (put $\Delta_{-1} = 0$).

Solutions (11) and (12) are equivalent. A proof by induction on j of this fact runs as follows. For $j = 1$, clearly $w_1 = d_1(0)$. For $j > 1$, it is sufficient to prove that

$$d_{j-1}(\delta_{j-1}) + w_j = d_j(0). \tag{13}$$

Noting that $\lambda \int_{\Delta_{j-2}} \Delta_j G\ c(t)dt = \rho_j - \rho_{j-2}$, that $\sum_{i=1} j - 1 t_i = d_{j-1}(\delta_{j-1}) + \Delta_{j-1}$, and using (11) for $d_{j-1}(\delta_{j-1})$, the left hand side of (13) reduces to

$$\frac{\lambda G\ c(\Delta_{j-1})E[S_j\ 2]}{2(1 - \rho_{j-1})(1 - \rho_j)} + \frac{\lambda E[\overline{S}_{j-1}\ 2]}{2(1 - \rho_{j-1})(1 - \rho_j)} + \frac{\rho_j \Delta_{j-1}}{1 - \rho_j}. \tag{14}$$

Simply evaluating the defining intergrals involved, one easily shows that

$$E[\overline{S}_j\ 2] - E[\overline{S}_{j-1}\ 2] - 2\Delta_{j-1} \int_{\Delta_{j-1}} \Delta_j G\ c(t)dt = G\ c(\Delta_{j-1})E[S_j\ 2]. \tag{15}$$

Using (15) in (14), and (11) for $d_j(0)$, the result now follows.

5 Discussion

Our original aim was to study the more general model with priority arrivals, that is, with K classes of jobs, class k jobs arriving at queue k according to a Poisson process with rate λ'_k, $k = 1, 2, \ldots, K$. Class k jobs then have service demand S_{kl} in queue l, and a fraction $q_{k,l+1}$ proceeds from queue l to queue $l + 1$, the rest leaving the system. The scheduling discipline is still preemptive resume. Mean waiting times for this model have been derived in [3] for QA, and in [6] for BF. The method used in this paper is not appropriate for this extended model (denoted by M as before).

Again, one could consider the modified model M_j with queues $1, 2, \ldots, j$ replaced by one queue J. In this case, the service time in queue J depends on the job class, but this can be handled by using an appropriate distribution derived from $\lambda'_1, \lambda'_2, \ldots, \lambda'_j$ and the $S_{kl}, k = 1, 2, \ldots, j, l = k, k + 1, \ldots, j$. However, we would now have to make two corrections to find the LST of the residence time of a class k job in queues $k, k + 1, \ldots, j$.

First, we cannot allow FCFS behavior in queue J, for even if a job, say of class k', arrives ahead of a tagged job A of class k, it is only entitled to the service it would have had in M up to and including queue $j - 1$, if $k' < k$, and the job would not have reached queue k in M before A arrives. Only taking together queues $1, 2, \ldots, j - 1$ does not help much, for then one would have to find out which jobs do get service in queue j ahead of A.

Second, as before we have to account for busy periods induced by arrivals to queues $1, 2, \ldots, j - 1$. However, these busy periods are now made up of two kinds of jobs. First, jobs that arrive 'ahead' of A (by this we mean jobs arriving in queues $k', k' + 1, \ldots, j$ while A is in one of the queues $k, k + 1, \ldots, k' - 1, k < k'$); these are entitled to service up to and including queue j before A's service in queue j. While A progresses, their arrival rate decreases because less classes are involved.

And second, jobs that arrive 'behind' A (that is, arrive in queues $1, 2, \ldots, k'$ while A is in one of the queues $k', k' + 1, \ldots, j$); these are entitled to service up to and including queue $j - 1$ before A's service in queue j. While A progresses, jobs of more classes are of this kind.

It seems very difficult to account for these corrections. Even if it were possible, it would require considering many queues at the same time, making the method much less elegant in this case.

Another extension is to introduce a preemption-distance D, with class k only preempting class l, if $k \leq l - D$. It seems one could deal with this extension by means of a straightforward combination of our treatment and the one in [7].

References

[1] T. Brown, "Determination of the Conditional Response for Quantum Allocation Algorithms," *Journal of the ACM*, Vol. 29, 448-460, 1982.

[2] R.B. Cooper, *Introduction to Queueing Theory*, The MacMillan Company, New York, 1972.

[3] D.H.J. Epema, "Mean Waiting Times in a General Feedback Queue with Priorities," *PERFORMANCE '90 - Proceedings of the 14th IFIP WG 7.3 International Symposium on Computer Performance Modelling, Measurement and Evaluation*, P.J.B. King, I. Mitrani and R.J. Pooley (eds.), Edinburgh, Scotland, 221-235, 1990 (also to appear in *Performance Evaluation*, 1991).

[4] L. Kleinrock, *Queueing Systems, Volume I: Theory*, John Wiley & Sons, New York, 1975.

[5] L. Kleinrock, *Queueing Systems, Volume II: Computer Applications*, John Wiley & Sons, New York, 1976.

[6] M. Paterok and O. Fischer, "Feedback Queues with Preemption-Distance Priorities," *ACM Sigmetrics Performance Evaluation Review and Performance '89*, Vol. 17, 136-145, 1989.

[7] L.E. Schrage, "The Queue M/G/1 with Feedback to Lower Priority Queues," *Management Science*, Vol. 13, 466-474, 1967.

[8] B. Simon, "Priority Queues with Feedback," *Journal of the ACM*, Vol. 31, 134-149, 1984.

[9] R.W. Wolff, "Time Sharing With Priorities," *SIAM Journal of Applied Mathematics*, Vol. 19, 566-574, 1970.

Approximate Analysis of a G/G/c/PR Queue[*]

Nasreddine M Tabet-Aouel
Demetres D Kouvatsos

Computer Systems Modelling Research Group
University of Bradford, Bradford
United Kingdom

Abstract

The principle of maximum entropy (ME) is used to analyse a stable G/G/c/PR queue with general interarrival and service times, c (≥ 2) parallel servers and R (≥ 2) priority classes under preemptive resume (PR) scheduling. The form of the joint state probability is characterised, subject to the existence of constraint information involving normalisation, marginal mean queue lengths and the probabilities of having a least number j of busy servers with jobs of class r. For $R = 2$, new closed-form approximations for the marginal state probabilities (per class), as explicit functions of the above constraints, are derived. Moreover, these ME solutions are also applied, in conjunction with the method of class aggregation, to the case of an arbitrary number of classes R (> 2). To illustrate the utility of the ME solutions the Generalised Exponential (GE) model is used to approximate general distributions with known first two moments, and exact as well as approximate stochastic analysis is carried out for estimating the associated constraints of interest. Numerical examples illustrate the accuracy of the proposed approximations in relation to simulations involving different interarrival and service time distributions per class. Concluding remarks and comments on the extension of the work to the analysis of general queueing networks are included.

1 Introduction

The G/G/c/PR queue with general interarrival and service time distributions, c (≥ 2) parallel servers and R (≥ 2) priority classes under preemptive resume (PR) scheduling is a very important building block for the analysis of complex queueing network models of computer systems and store-and-forward packet-switched data networks with parallel processors where jobs and packets, respectively, are subject to the PR rule. In principle, the stochastic analysis of this queue in isolation depends on the choice of the general (G-type) priority interarrival and service time distributions and is characterised by measures such as queue length and response time (c.f. [1]).

Unfortunately, due to the complexity involved, very few exact or approximate results can be found in the literature (c.f. [2–5]). Jaiswal [1, p.217] has described

[*]This work is sponsored by the Science and Engineering Research Council (SERC), UK, under grant GR/F29271

the analysis of this type of queues as 'an extremely complicated mathematical problem' and has suggested that the main objective should be directed towards the search for simpler methods for studying such a non-Markovian process which requires more than one supplementary variable (when adopting the classical supplementary variable technique). Furthermore, to the knowledge of the authors, no analytic closed-form approximation for the joint and marginal queue length distribution of a G/G/c/PR queue has so far been proposed in the literature. For the case of a stable M/M/c/PR queue, Mitrani and King [2] derived analytic expressions for the generating functions of the joint and marginal queue length distributions which must be evaluated numerically for each set of model parameters. The solution of the equations is difficult and, as the authors note, subject to numerical instability. For more than two priority classes, they suggested a heuristic in which a two class system is created consisting of the class-r of interest at the second priority level and the aggregation of all higher priority classes $\{1, 2, \ldots, r-1\}$ at the first level. The system is then evaluated as a two-priority class system. This approximation, however, suffers from potential sources of errors as it implies (i) a FCFS. (first-come-first-served) policy amongst classes $1, 2, \ldots, r-1$ (instead of PR) and (ii) an exponentially distributed service time for the aggregate class (instead of hyperexponential). Intuitively, the errors caused by the second source are likely to be much larger than those caused by the first (c.f. [2]).

This paper focusses on the approximate analysis of a stable G/G/c/PR queue. The form of the joint state probability is characterised, based on the principle of maximum entropy (ME) (c.f. Jaynes [6]), subject to constraint information involving marginal mean queue lengths $\{\langle n_r \rangle, r = 1, \ldots, R\}$ and statistics $\{U_{rj}, r = 1, \ldots, R; j = 1, \ldots, c\}$ defining the probabilities of having a minimum of j jobs of class-r in service (i.e. j or more servers are busy with class-r jobs). Moreover, for $R = 2$, the marginal state probabilities are analytically derived in terms of the above constraints, while for $R > 2$ the class aggregation method is applied for creating a two-class system (c.f. Mitrani and King [2]). The implementation of the ME solutions requires the *a priori* analytic estimation of constraints $\{\langle n_r \rangle\}$ and $\{U_{rj}\}$. To this end, exact and approximate stochastic analysis is carried out based on the use of the GE distributional model of the form

$$\phi(t) = 1 - \frac{2}{C+1} \exp\left[\frac{-2\nu t}{C+1}\right], \qquad t \geq 0, \tag{1}$$

(where $1/\nu$ and C are the mean value and squared coefficient of variation (SCV), respectively) in approximating the interarrival and service time distributions with known first two moments (c.f. [7–10]).

The principle of ME, subject to normalisation and marginal constraints $\{\langle n_r \rangle\}$ and $\{U_{rj}\}$, is used in Section 2 to characterise the form of the solution of the joint state probability of a stable G/G/c/PR queue. Section 3 contains ME solutions for the case $R = 2$ together with approximate GE-type formula for $\{\langle n_r \rangle\}$ and $\{U_{rj}\}$, respectively. These results are used in turn in Section 4 for the approximate analysis of a stable G/G/c/PR queue with $R > 2$ classes within the context of the class aggregation method. Section 5 presents numerical results involving ME solutions and simulations using different underlying probability distributions. Conclusions and further comments on the implication of the results into the approximate analysis of complex queueing networks follow in the last section.

2   ME formalism and the G/G/c/PR queue

Consider a stable G/G/c/PR queue with R classes of jobs where the priority level is in decreasing order of the class index (i.e. class-r job has priority over class-s if $r < s$). Jobs arrive according to an arbitrary distribution, G, with arrival rate λ_r and SCV C_{ar}, and are served by c parallel homogeneous servers with service rate μ_r and SCV C_{sr}, $r = 1, \ldots, R$. Note that the case '$c = 1$' has been fully investigated in [9, 10].

Let at any given time the state of the system be described by a vector $\underline{n} = (n_1, n_2, \ldots, n_R)$, where n_r, $r = 1, 2, \ldots, R$ is the number of class-r jobs in the queue (waiting for or receiving service) such that $n_r \in [0, +\infty)$ and Q be the state space of the system. Moreover, $P(\underline{n})$ is the joint state probability and $P_r(\underline{n_r})$ is the marginal state probability of class-r, $r = 1, 2, \ldots, R$. Suppose all that is known about the state probability $\{P(\underline{n})\}$ is that the following mean value constraints exist:

(a) Normalisation (norm),
$$\sum_{\underline{n} \in Q} P(\underline{n}) = 1. \tag{2}$$

(b) Probabilities, $\{U_{rj}, r = 1, \ldots, R; j = 1, \ldots, c\}$, $U_{rj} \in (0, 1)$,
$$\sum_{\underline{n} \in Q} h_{rj}(\underline{n}) P(\underline{n}) = U_{rj}, \qquad r = 1, \ldots, R; j = 1, \ldots, c, \tag{3}$$

where
$$h_{rj}(\underline{n}) = \begin{cases} 1, & \text{if } n_r \geq j \text{ and } j \leq c - \sum_{l=1}^{r-1} n_l, \\ 0, & \text{otherwise,} \end{cases}$$

(c) Mean queue lengths, $\{\langle n_r \rangle, r = 1, \ldots, R\}$, $0 < \langle n_r \rangle < +\infty$,
$$\sum_{\underline{n} \in Q} n_r P(\underline{n}) = \langle n_r \rangle, \qquad r = 1, \ldots, R \tag{4}$$

Note that the inclusion of constraints (3) and (4) is motivated by the type of constraints applied in the ME analysis at equilibrium of both the G/G/1/PR queue (c.f. Kouvatsos and Tabet-Aouel [9]) and the single class G/G/c/FCFS queue (c.f. Kouvatsos and Almond [11]).

The principle of ME (c.f. Jaynes [6]) states that the least biased distribution supported by the constraints (2) – (4) is the one that maximises the entropy functional, $H(p)$, namely
$$H(P) = \max\left\{ -\sum_{\underline{n} \in Q} P(\underline{n}) \log[P(\underline{n})] \right\}, \tag{5}$$

subject to constraints (2) – (4).

The maximisation of (5), subject to constraints (2) – (4), can be carried out by using the Lagrange method of undetermined multipliers leading to the following general solution of the form
$$P(\underline{n}) = \frac{1}{Z} \prod_{r=1}^{R} \left\{ \prod_{j=1}^{c} g_{rj}^{h_{rj}(n)} \right\} X_r^{n_r}, \tag{6}$$

where $\{g_{rj}\}$ and $\{X_r\}$, $r = 1,\ldots,R$ and $j = 1,\ldots,c$ are the Lagrange coefficients corresponding to constraints (3) and (4), respectively, and Z is the normalising constant associated with constraint (2).

In general, the analytic estimation of the Lagrange coefficients $\{X_r\}$, $\{g_{rj}\}$ $r = 1,\ldots,R$ and $j = 1,\ldots,c$ in terms of the constraints $\{U_{rj}\}$ and $\{\langle n_r \rangle\}$ has been found very difficult to obtain directly from the ME solution (6). However, for the special case $R = 2$, "exact" ME analysis is possible and the main analytic results are given next.

3 The G/G/c/PR queue with $R = 2$: Analytic results

3.1 ME solution

The ME solution (6) and the Lagrangian coefficients are determined by the following theorem:

Theorem 1. The joint maximum entropy solution, $p(\underline{n})$, of a stable G/G/c/PR queue with $R = 2$ priority classes, subject to prior information (Norm, U_{rj}, $\langle n_r \rangle$, $r = 1,2$ and $j = 1,\ldots,c$) is given by

$$
P(n_1, n_2) = \begin{cases}
\frac{1}{Z} = P(\underline{0}), & \\[2mm]
P(\underline{0})G_{1n_1}G_{2n_2}X_1^{n_1}X_2^{n_2}, & n_1 = 0,\ldots,c-1; n_2 = 0,\ldots,c-n_1-1, \\[2mm]
P(\underline{0})G_{1n_1}G_{2c-n_1}X_1^{n_1}X_2^{n_2}, & n_1 = 0,\ldots,c-1; n_2 \geq c-n_1, \\[2mm]
P(\underline{0})G_{1c}X_1^{n_1}X_2^{n_2}, & n_1 \geq c; n_2 \geq 0,
\end{cases}
\tag{7}
$$

where $G_{rj} = \prod_{i=0}^{j} g_{ri}$ and $G_{r0} = 1$, $r = 1,\ldots,R$ and $j = 1,\ldots,c$ and are given by

$$
G_{1j} = \begin{cases}
\dfrac{(1-X_1)(1-X_2)U_{1c}}{P(\underline{0})X_1^c}, & \\[4mm]
\dfrac{(1-X_2)(U_{1j}-U_{1j+1})}{\left[(1-X_2)\sum\limits_{n_2=0}^{c-1-j}P(0,n_2)+P(0,c-j)\right]X_1^j}, & j = 1,\ldots,c-1,
\end{cases}
\tag{8}
$$

$$
G_{2j} = \begin{cases}
\dfrac{(1-X_2)U_{2c}}{P(\underline{0})X_2^c}, & \\[4mm]
\dfrac{U_{2j}-U_{2j+1}-(U_{1c-j}-U_{1c-j+1})+\sum\limits_{n_2=0}^{j-1}P(c-j,n_2)}{\left[X_2^j\sum\limits_{n_2=0}^{c-1-j}P(n_1,0)\right]}, & j = 1,\ldots,c-1,
\end{cases}
\tag{9}
$$

$$X_1 = \frac{\langle n_1 \rangle - \sum_{j=1}^{c} U_{1j}}{\langle n_1 \rangle - \sum_{j=1}^{c-1} U_{1j}}, \tag{10}$$

$$X_2 = \frac{\langle n_2 \rangle - \sum_{j=1}^{c} U_{2j}}{\langle n_2 \rangle - \left[\sum_{j=1}^{c-1} U_{2j} + \sum_{n_1=1}^{c-1} \sum_{n_2=0}^{c-1-n_1} P(n_1, n_2) - U_{11} \right]}, \tag{11}$$

$$Z = \frac{1}{P(\underline{0})}. \tag{12}$$

Proof: The ME solution (7) is established directly by using equations (2) and (6). By considering constraints (3) and (4), after some manipulation, expressions (8) – (11) are derived. Equation (12) follows directly from the definition of $h_{rj}(\underline{n})$ of equation (3). Q.E.D.

From the ME solution (7) of Theorem 1, the marginal probability $P_r(n_r)$ of class-r, $r = 1, 2$, is determined by the following corollary:

Corollary 1. The marginal ME solution $P_r(n_r)$ of class-r jobs, $r = 1, 2$ ($R = 2$) of a stable G/G/c/PR queue, subject to prior information (Norm, U_{rj}, $\langle n_r \rangle$, $r = 1, 2$ and $j = 1, \ldots, c$), is given by

$$P_r(n_r) = \begin{cases} \hat{U}_{rn_r} - \hat{U}_{rn_r+1}, & n_r = 0, \ldots, c-1 \\ \hat{U}_{rc}(1 - X_r)X_r^{n_r-c}, & \text{for } n_r \geq c, \end{cases} \tag{13}$$

where

$$\hat{U}_{rj} = \begin{cases} U_{1j}, & r = 1, \\ U_{r1} + X_r^j U_{r-1c} - \left[\sum_{n_2=1}^{j=1} \sum_{n_1=0}^{c-n_2} P(n_1, n_2) + \sum_{k=1}^{j-2} X_r^k \sum_{n_2=1}^{j-1-k} P(c - n_2, n_2) \right], & r = 2. \end{cases} \tag{14}$$

Proof: The proof is conducted by evaluating the marginal probabilities

$$P_r(n_r) = \sum_{\underline{n} \in Q \ \& \ n_r = k} P(\underline{n}),$$

where $n_r = k \in [0, +\infty)$ and $P(\underline{n})$ is the joint probability distribution (7). Note that, as expected, the ME marginal probability, $P_1(n_1)$, is identical to the ME queue length distribution of an ordinary G/G/c queue (c.f. [11]). However, the ME marginal probabilities, $P_2(n_2)$, (obtained by the above summation) have a different analytic form, namely

$$P_2(n_2) = \begin{cases} \displaystyle\sum_{n_1=0}^{c-1} P(n_1,0) + (1-X_2)U_{1c}, & n_2 = 0 \\[3mm] \left[X_2^{\frac{1}{m_2}} \displaystyle\sum_{n_1=0}^{c-m_2} P(n_1,m_2) + \displaystyle\sum_{j=1}^{m_2-1} \frac{P(c-j,j)}{X_2^j} + (1-X_2)U_{1c} \right] X_2^{n_2}, & n_2 \geq 1, \end{cases}$$

$$\tag{15}$$

where

$$m_2 = \begin{cases} n_2, & \text{for } n_2 < c, \\ c, & \text{for } n_2 \geq c. \end{cases}$$

By using some variable-changes and after simple calculations, equation (13) follows. Q.E.D.

Note that the Lagrange multipliers $\{X_r, r = 1,2\}$ may be written in the following form

$$X_r = \frac{\langle n_r \rangle - \displaystyle\sum_{j=1}^{c} \widehat{U}_{rj}}{\langle n_r \rangle - \displaystyle\sum_{j=1}^{c-1} \widehat{U}_{rj}}, \qquad r = 1,2. \tag{16}$$

It is interesting to point out that the marginal ME solution (13) for class-r jobs is equivalent to the ME queue length distribution of an ordinary single-class G/G/c queue at equilibrium (c.f. [11]), subject to the "new" utilisations, $\{\widehat{U}_{rj}\}$ and the "old" mean queue lengths, $\{\langle n_r \rangle\}$, constraints. In this case, "$\widehat{U}_{rj}$" represents the perceived probability to have at least j jobs of class-r in service of a dedicated G/G/c queue with (i) inflated service times to take into account the effect of high priority jobs and (ii) a queue length distribution corresponding exactly to the marginal ME solution $P_r(n_r)$ (given by equation (13)). Moreover, to establish an admissible ME approximation, when improper interarrival and/or service time distributions are involved (with SCVs less than 1, e.g. GEs), the following relation should be satisfied

$$\langle n_r \rangle > \sum_{j=1}^{c} \widehat{U}_{rj}, \tag{17}$$

or equivalently, $X_r > 0$.

This condition is analogous to the one seen in the ME analysis of an ordinary G/G/1 queue (c.f. [7,8]), where the mean queue length, $\langle n \rangle$, should be strictly greater than the utilisation, ρ (i.e. $\langle n \rangle > \rho$).

Remark: It is important to note that the ME solutions (7) and (13) involve the joint probabilities $P(n_1, n_2)$ with $n_1 = 0, \ldots, c-1$ and $n_2 = 0, \ldots, c - n_1$. Although these probabilities (except $P(0,0)$) do not explicitly correspond to constraints (2) – (4), they can be obtained directly from the normalisation constant "$1/Z$" (or the idle state probabilities $P(\underline{0})$). This is because the sub-space $Q^* = \{(n_1, n_2)/n_1 < c, n_2 \leq c - n_1 \text{ and } n_1 + n_2 > 0\}$ represents the subset of Q where the priority discipline cannot be applied. This is due to the fact that no congestion (queueing) occurs in these cases. Jobs in these circumstances are served upon their arrival instant and the scheduling is an infinite server discipline. To that extent the joint probabilities $P(n_1, n_2)$, with (n_1, n_2) belonging to the sub-space Q^*, can be given by

$$P(\underline{n}) = P(\underline{0}) \prod_{r=1}^{R} f_r(n_r), \qquad (18)$$

where $f_r(n_r)$ is the unnormalised marginal ME solution of class-r in an ordinary $G/G/\infty$ queue, the solution of which can be found in [11].

3.2 GE-type approximation constraints

This subsection focusses on the GE/GE/c/PR queue with $R = 2$ and carries out exact and approximate analyses in order to establish analytic approximations for the statistics $P(0,0)$, $\{\langle n_r \rangle\}$ and $\{U_{rj}\}$, $r = 1, 2; j = 1, \ldots, c$ of a stable $G/G/c/PR$ queue.

3.2.1 Approximation of $P(0,0)$

Probability $P(0,0)$ is the idle state probability of a stable $G/G/c/PR$ queue with $R = 2$ classes of jobs and corresponds directly to the normalisation constraint (2). The estimation of this probability is essential in computing joint probabilities $\{P(n_1, n_2)\}$, $(n_1, n_2) \in Q^*$. Because in the GE/GE/c/PR queue there is no creation or destruction of service, $P(0,0)$ is accurately approximated by the idle state probability of an ordinary single-class GE/GE/c queue where the two classes are coalesced into a single one. Namely (c.f. Kouvatsos and Almond [11])

$$P(0,0) = \left[1 + \sum_{n=1}^{c-1} G_n + \frac{G_c}{1 - X} \right]^{-1}, \qquad (19)$$

where

$$G_n = \prod_{k=1}^{n} g_k, \qquad (20)$$

and

$$g_k = \begin{cases} \dfrac{\overline{\lambda}_2(\overline{C}_{s2} + 1) + \overline{\mu}_2(k-1)(\overline{C}_{a2} - 1)}{k\overline{\mu}_2(\overline{C}_{a2} + \overline{C}_{s2})}, & k = 1, \ldots, c-1 \\[4mm] \dfrac{\overline{\lambda}_2(\overline{C}_{s2} + 1) + \overline{\mu}_2(k-1)(\overline{C}_{a2} - 1)}{\overline{\lambda}_2(\overline{C}_{s2} - 1) + k\overline{\mu}_2(\overline{C}_{a2} + 1)} \cdot \dfrac{2}{\overline{C}_{s2} + 1}, & k = c \end{cases} \qquad (21)$$

and

$$X = \frac{\overline{\lambda}_2(\overline{C}_{s2} + 1) + \overline{\mu}_2 c(\overline{C}_{a2} - 1)}{\overline{\lambda}_2(\overline{C}_{s2} - 1) + c\overline{\mu}_2(\overline{C}_{a2} + 1)} \qquad (22)$$

and $\overline{\lambda}_2$, $\overline{\mu}_2$, $\overline{C}_{a2}$, $\overline{C}_{s2}$ are the parameters of the coalesced class (c.f. Appendix).

3.2.2 Approximation of $\{U_{rj}\}$

Constraints $\{U_{rj}, r = 1, 2; j = 1, \ldots, c\}$ are by definition the steady state probabilities to have a minimum number of jobs of class-r in service. Furthermore, under PR discipline, class-1 jobs are not at all influenced by the presence of low-priority jobs (class-2). Therefore, it is clearly evident that the quantities $U_{1j}, j = 1, \ldots, c$ are identical to the ones in an ordinary G/G/c queue. For the special case of GE interarrival and service-time distribution, U_{1j} are given by (c.f. Kouvatsos and Almond [11])

$$
U_{1j} = \begin{cases} A\left[\displaystyle\sum_{n=j}^{c-1} \Delta_n + \frac{\Delta_c}{1-\omega}\right], & j = 0, \ldots, c-1, \\[4mm] A\dfrac{\Delta_c}{1-\omega}, & j = c, \end{cases} \tag{23}
$$

where

$$
\Delta_n = \prod_{k=1}^{n} \delta_k, \tag{24}
$$

and

$$
\delta_k = \begin{cases} \dfrac{\lambda_1(C_{s1}+1) + \mu_1(k-1)(C_{a1}-1)}{k\mu_1(C_{a1}+C_{s1})}, & k = 1, \ldots, c-1, \\[4mm] \dfrac{\lambda_1(C_{s1}+1) + \mu_1(k-1)(C_{a1}-1)}{\lambda_1(C_{s1}-1) + k\mu_1(C_{a1}+1)} \cdot \dfrac{2}{C_{s1}+1}, & k = c, \end{cases} \tag{25}
$$

and

$$
\omega = \frac{\lambda_1(C_{s1}+1) + \mu_1 c(C_{a1}-1)}{\lambda_1(C_{s1}-1) + c\mu_1(C_{a1}+1)}, \tag{26}
$$

and

$$
A = \left[1 + \sum_{n=1}^{c-1} \Delta_n + \frac{\Delta_c}{1-\omega}\right]^{-1}. \tag{27}
$$

For $r = 2$, U_{2j} can be determined directly by simple probabilistic arguments, based on the knowledge of $U_{1l}, l = 1, \ldots, c$ and some joint probabilities $P(n_1, n_2)$ with the states (n_1, n_2) belonging to the sub-space Q^*. They are given by the following corollary:

Corollary 2: The steady state probabilities $U_{2j}, j = 1, \ldots, c$, defining the minimum number of j jobs of class-2 in service in a stable G/G/c/PR queue with $R = 2$ classes of jobs is given by the following general form

$$
U_{2j} = \begin{cases} 1, & j = 0, \\[3mm] U_{2j+1} + U_{1c-j} - U_{1c-j+1} - \displaystyle\sum_{n_2=0}^{j-1} P(c-j, n_2) + \sum_{k=0}^{c-j-1} P(k, j), & 0 < j < c, \\[4mm] 1 - U_{11} - \displaystyle\sum_{k=0}^{c-1} P(0, k), & j = c. \end{cases} \tag{28}
$$

116

Proof: The evaluation of U_{2j}, $j = 1, \ldots, c$ is done in decreasing order of the index j. In other words, U_{2c} must be the first one to be computed and then comes the evaluation of U_{2c-1}, U_{2c-2}, $\ldots$, U_{21}. In the case of PR discipline and 2 classes of jobs, it follows that

$$
\begin{aligned}
U_{2c} &= \sum_{n_2=0}^{\infty} P(0, n_2) \\
&= \sum_{n_2=0}^{\infty} P(0, n_2) - \sum_{n_2=0}^{c-1} P(0, n_2) \\
&= P_1(0) - \sum_{n_2=0}^{c-1} P(0, n_2).
\end{aligned}
$$

Equation (28) for $j = c$ follows after substituting U_{11} by $1 - P_1(0)$.

To evaluate U_{2j} for $j < c$, it is assumed that U_{2j+1} has already been determined since the evaluation procedure is done in decreasing order of the index j. Therefore, U_{2j} may be interpreted as the sum of two probabilities, namely

$$
\begin{aligned}
U_{2j} = {}&\text{Prob } [j \text{ jobs of class-2 in service}] + \\
&\text{Prob } [\text{at least } j + 1 \text{ jobs of class-2 in service}],
\end{aligned}
$$

which can be analytically written as

$$
U_{2j} = U_{2j+1} + \sum_{k=0}^{c-j-1} P(k, j) + \sum_{n_2=j}^{\infty} P(c - j, n_2),
$$

and after simple probabilistic manipulations, equation (28) with $j < c$ follows. Q.E.D.

3.2.3 Approximation of $\langle n_r \rangle$

The marginal mean queue lengths, $\{\langle n_r \rangle\}$, $r = 1, \ldots, R \ (\geq 2)$, can be computed by modifying the approximate method proposed by Buzen and Bondi [3,4] to estimate the marginal response times $\{W_r, r = 1, \ldots, R\}$ of a stable GE/GE/c/PR queue and applying Little's law. The approximation can be expressed by

$$
W_r = \frac{\overline{\lambda}_r \overline{W}_r - \overline{\lambda}_{r-1} \overline{W}_{r-1}}{\lambda_r}, \qquad r = 1, \ldots, R \tag{29}
$$

where $\overline{\lambda}_r$ is the overall arrival rate of the first r classes (see Appendix) and $\overline{W}_r$ is the overall mean response time of r-class c-server system under PR rule, namely

$$
\overline{W}_r = \overline{Q}_r + 1/\overline{\mu}_r, \tag{30}
$$

where $\overline{\mu}_r$ is the aggregate service rate over classes $1, 2, \ldots, r$ and $\overline{Q}_r$ is the overall queueing (or waiting) time of class-r c-server system denoted in general by a function $Q(d, \underline{\lambda}_r, \underline{C}_{ar}, \underline{\mu}_r, \underline{C}_{sr}, c)$ where scheduling discipline d is either FCFS or PR, and $\underline{\lambda}_r$, $\underline{C}_{ar}$, $\underline{\mu}_r$, and $\underline{C}_{sr}$ are r-vectors of the corresponding interarrival and service time parameters and is expressed by

$$Q(PR, \underline{\lambda}_r, \underline{C}_{ar}, \underline{\mu}_r, \underline{C}_{sr}, c) \approx Q(PR, \underline{\lambda}_r, \underline{C}_{ar}, \underline{\mu}_r, \underline{C}_{sr}, 1) \times \frac{Q(FCFS, \overline{\lambda}_r, \overline{C}_{ar}, \overline{\mu}_r, \overline{C}_{sr}, c)}{Q(FCFS, \overline{\lambda}_r, \overline{C}_{ar}, \overline{\mu}_r, \overline{C}_{sr}, 1)}$$
(31)

where the ratio in equation (31) is that of a single aggregate class of waiting times with the given parameters. The mean waiting times of the right hand side of (31) can be easily evaluated by using earlier results from the ME analysis of (i) the stable GE/GE/1/PR queue (c.f. [9,10]) and (ii) the ordinary but stable GE/GE/1 and GE/GE/c queues (c.f. [8,11]), respectively. Further details of determining the approximations, together with numerical tests, can be seen in [5].

4 The G/G/c/PR queue with $R > 2$

As mentioned earlier, for $r > 2$, the analytic derivation of the Lagrange coefficients directly from the ME solution (6) is very tedious and computationally expensive (due to the large number of joint probabilities involved corresponding to the non-congestion states, $\{S\}, S \in Q^*$). It is worth emphasising that the analytic expressions of the Lagrange coefficients and the marginal queue length distributions (c.f. equations (8) — (11)) are still valid for $r = 1, 2$ when the number of classes R exceeds 2. This is due to the fact that in a G/G/c/PR queue, lower priority class jobs ($r > 2$) do not have any influence whatsoever on the performance of jobs belonging to class-1 and class-2.

The problem can be tackled heuristically by employing the aggregation technique (c.f. [2]) together with the use of earlier results of Section 3. Namely, the class-r ($r > 2$) characteristics are obtained, firstly by ignoring the presence of low-priority jobs, since these jobs do not affect the queueing time of class-r jobs and secondly, by using the 2-class results where high-priority classes ($s < r$) are amalgamated into a single coalesced-class which constitutes the class-1 (in the 2-class system) and class-2 represents the class-r in question. In particular, the aggregation technique is used only to approximate the Lagrange coefficients, $\{X_r\}$, $\{g_{rj}\}$ and statistics "$\hat{U}_{rj}$" (i.e. the perceived probability to have j jobs of class-r in service as they are served by c dedicated homogeneous servers with inflated service times to take into account the effect of high priority jobs). This is due to the fact that the form of the ME solution is known in a general case ($R > 2$) and given by equations (6) and (13) for the joint and marginal queue length distributions, respectively.

Note that this approximation attempts to capture the variability error observed in Mitrani and King [2], since a GE-type distribution is used to represent the service-time of the aggregate class of jobs.

5 Numerical results

This section presents typical numerical results on stable G/G/c/PR queues with $R = 2, 3$ and 4 classes of jobs in order to demonstrate the credibility of the maximum entropy approximations of the previous sections. The ME solutions are based on marginal GE-type constraints (3) – (4) and are validated against simulation experiments (denoted by SIM) using different forms of general distributions with

		Class-r				
		$r = 1$			$r = 2$	
λ_r		2			0.75	
C_{ar}		100			95	
μ_r		2			1	
C_{sr}		21			78	
		SIM		ME	SIM		ME
		$H_2\,(\kappa = 2)$	GE		$H_2\,(\kappa = 2)$	GE	
	0	0.441	0.458	0.458	0.151	0.141	0.111
	1	0.103	0.083	0.083	0.041	0.080	0.039
	2	0.043	0.003	0.004	0.010	0.002	0.002
$P_r(n_r)$	3	0.028	0.004	0.004	0.007	0.002	0.002
	4	0.020	0.004	0.004	0.005	0.002	0.002
	5	0.015	0.004	0.004	0.004	0.003	0.002

Table 1: Comparison of ME marginal probabilities, $\{P_r(n_r)\}$, $r = 1, 2$ and $n_r = 0, \ldots, 5$ of a G/G/2/PR queue with $R = 2$ and $c = 2$ against simulations (SIM) involving interarrival and service time distributions of type H_2 ($\kappa = 2$) and GE

known first two moments, including Deterministic (D), Uniform ($U(a, b)$), Erlang-2 (E_2), Exponential (M), balanced Hyperexponential-2 ($H_2\,(\kappa = 2)$) and GE distributional models, where κ is a real tuning parameter used to completely define an H_2 distribution with known first two moments (c.f. [7–11]) and (a, b) is a real interval in which a continuous uniform random variable is defined. The simulation results are produced at 95% confidence intervals by making use of the queueing network analysis package QNAP-2 [12].

Table 1 exhibits the marginal probabilities $\{P_r(n_r)\}$, $r = 1, 2$ of a 2-class G/G/2/PR with H_2 ($\kappa = 2$) and GE interarrival and service time distributions with SCVs ranging from 21 to 100. It is observed that the ME solution is consistently comparable to that of simulation models with absolute derivation less than 0.5. In particular, the results show that the ME approximation improves, as expected, when the simulation models employ the GE at $\kappa \to +\infty$ (instead of H_2 ($\kappa = 2$)) distribution. Note that, as it was conjectured in [5], the GE provides performance bounds over a wide range of distributions on the marginal mean response times. The credibility of the ME approximation for the G/G/4/PR queue is further illustrated in Tables 2 and 3 for two and three priority classes, respectively, by considering heterogeneous types of interarrival and service time distributions per class including D, U, E_2, M and GE distributional models. These example queues are strikingly more complex than those considered in Table 1 and show the versatility of the ME solution and the GE-type formulae which can be utilised even for distributions with SCVs less than one given that the condition expressed by equation (17) is satisfied. Note that in Table 3, the class aggregation technique is used (c.f. Section 4) to generate the marginal probabilities of class-3. Finally, in Table 4, four priority classes are considered in order to assess the accuracy of the G/G/3/PR ME

		Class-r		
		$r = 1$		$r = 2$	
λ_r		1		3	
C_{ar}		4 (GE)		2 (GE)	
μ_r		2		1	
C_{sr}		0.5 (E_2)		1/3 ($U(0, 2/\mu_2)$)	
		SIM	ME	SIM	ME
	0	0.750	0.748	0.035	0.045
	1	0.122	0.124	0.067	0.074
	2	0.054	0.052	0.081	0.079
$P_r(n_r)$	3	0.028	0.016	0.079	0.068
	4	0.015	0.017	0.072	0.074
	5	0.010	0.011	0.067	0.066

Table 2: Comparison of ME marginal probabilities, $\{P_r(n_r)\}$, $r = 1, 2$ and $n_r = 0, \ldots, 5$ of a G/G/4/PR queue with $R = 2$ and $c = 4$ against simulations (SIM) involving interarrival and service time distributions

		Class-r					
		$r = 1$		$r = 2$		$r = 3$	
λ_r		1		3		1	
C_{ar}		4 (GE)		2 (GE)		5 (GE)	
μ_r		2		1		5	
C_{sr}		0.5 (E_2)		1/3 ($U(0, 2/\mu_2)$)		0 (D)	
		SIM	ME	SIM	ME	SIM	ME
	0	0.749	0.748	0.035	0.045	0.139	0.153
	1	0.123	0.124	0.066	0.074	0.075	0.028
	2	0.054	0.052	0.080	0.079	0.057	0.025
$P_r(n_r)$	3	0.027	0.026	0.079	0.068	0.041	0.024
	4	0.016	0.017	0.071	0.074	0.033	0.023
	5	0.010	0.011	0.066	0.066	0.028	0.022

Table 3: Comparison of ME marginal probabilities, $\{P_r(n_r)\}$, $r = 1, \ldots, 3$ and $n_r = 0, \ldots, 5$ of a G/G/4/PR queue with $R = 3$ and $c = 4$ against simulations (SIM) involving interarrival and service time distributions

		Class-r						
		$r=1$		$r=2$		$r=3$		$r=4$
λ_r		3		2		1		5
C_{ar}		25 (GE)		17 (GE)		3 (GE)		1 (GE)
μ_r		5		2		2		8
C_{sr}		35 (GE)		14 (GE)		7 (GE)		5 (GE)
		SIM	ME	SIM	ME	SIM	ME	SIM	ME
	0	0.662	0.668	0.404	0.400	0.341	0.339	0.089	0.086
	1	0.221	0.224	0.199	0.194	0.159	0.146	0.056	0.054
	2	0.085	0.085	0.088	0.106	0.063	0.075	0.021	0.030
$P_r(n_r)$	3	0.003	0.003	0.013	0.013	0.035	0.024	0.011	0.003
	4	0.003	0.003	0.013	0.012	0.029	0.023	0.009	0.003
	5	0.003	0.003	0.012	0.011	0.026	0.022	0.008	0.003

Table 4: Comparison of ME marginal probabilities, $\{P_r(n_r)\}$, $r = 1, \ldots, R$ and $n_r = 0, \ldots, 5$ of a G/G/3/PR queue with $R = 4$ and $c = 3$ against simulations (SIM) involving interarrival and service time distributions

solution for $c = 3$ and $R > 2$ number of classes. In essence, the results of classes 3 and 4 (obtained via class aggregation) show that the source of error caused by implying a FCFS rule amongst classes $1, 2, \ldots, r - 1$ (instead of PR) is less important than neglecting the effect of the variability of the aggregate service time distribution which is satisfactorily captured by the ME model via the GE-type formulae.

All the experiments carried out indicate that the mean queue lengths $\{\langle n_r \rangle\}$ and the probabilities $\{U_{rj}\}$, $r = 1, \ldots, R$ and $j = 1, \ldots, c$ are sufficient constraints enabling the ME solution to capture the shape of the entire queue length distribution per class with considerable accuracy.

6 Conclusions and further comments

The principle of ME, in conjunction with the method of class aggregation, is used to characterise approximately the forms of the joint and marginal state probabilities of a stable G/G/c/PR queue with c (≥ 2) parallel servers and R (≥ 2) priority classes, subject to constraint information involving the normalisation and statistics $\{\langle n_r \rangle\}$ and $\{U_{rj}\}$, $r = 1, \ldots, R$ and $j = 1, \ldots, c$. The ME solutions are implemented by utilising the GE distributional model in approximating general distributions with known first two moments. Numerical experiments illustrate the high level of accuracy and robustness of the ME approximations against simulations involving different interarrival and service time distributions per class.

The ME formalism can also be applied to determine a product-form approximation for a general queueing network with multiple parallel servers at each station i and R (≥ 2) classes of jobs under PR and FCFS scheduling rules, subject to normalisation and marginal constraints $\{\langle n_r \rangle\}$ and $\{U_{rj}\}$ for each station i. In this context, the new ME solution of a stable G/G/c/PR queue presented here may be used as a building block, in conjunction with appropriate flow formulae, for the approximate station-by-station decomposition analysis of the entire network. Work of this kind is the subject of current study.

Appendix

For the first r priority classes of jobs, $r \in [1, R]$, let $\overline{\lambda}_r$ be the overall arrival rate, $\overline{\mu}_r$ be the overall service rate, $\overline{C}_{ar}$ be the overall SCV of the interarrival-time, $\overline{C}_{sr}$ be the overall SCV of the service-time.

Clearly $\overline{\lambda}_r$ and $\overline{\mu}_r$ are given by

$$\overline{\lambda}_r = \sum_{s=1}^{r} \lambda_s, \qquad r = 1, \dots, R,$$

and (via the law of total expectations)

$$\overline{\mu}_r = \frac{\overline{\lambda}_r}{\sum_{s=1}^{r} \frac{\lambda_s}{\mu_s}},$$

respectively.

Moreover, $\overline{C}_{ar}$ is determined by the merging formula of r GE streams [13], namely

$$\overline{C}_{ar} = -1 + \left\{ \sum_{s=1}^{r} \frac{\lambda_s}{\overline{\lambda}_r} (C_{as} + 1)^{-1} \right\}^{-1},$$

and by applying the law of total expectations, $\overline{C}_{sr}$ is given by [13]

$$\overline{C}_{sr} = \frac{\overline{\mu}_r^2}{\overline{\lambda}_r} \left\{ \sum_{l=1}^{r} \frac{\lambda_l}{\mu_l^2} (C_{sl} + 1) \right\} - 1.$$

References

[1] Jaiswal, N.K. Priority queues, 1st Ed., Academic Press, New York, 1968.

[2] Mitrani, I. and King, P.J.B. Multiprocessor systems with pre-emptive priorities. Perf. Eval. 1981; 1: 118-125.

[3] Buzen, J.P. and Bondi, A.B. The response time of priority classes under pre-emptive resume in M/M/m queues. Oper. Res. 1983; 31(3): 456–465.

[4] Bondi, A.B. and Buzen, J.P. The response time of priority classes under pre-emptive resume in M/G/m queues. J. ACM 1984; 195–201.

[5] Tabet-Aouel, N. and Kouvatsos, D.D. On the approximation of the mean response times of priority classes in a stable G/G/c/PR queue. Tech. Rep. CS-21-90, University of Bradford, Bradford, 1990.

[6] Jaynes, E.T. Prior probabilities. IEEE Trans. Syst. Sci. Cyber SSC 1968; 4: 227–241.

[7] El-Affendi, M.A. and Kouvatsos, D.D. A maximum entropy analysis of the M/G/1 and G/M/1 queueing systems at equilibrium. Acta Info. 1983; 19: 339–355.

[8] Kouvatsos, D.D. A maximum entropy analysis of the G/G/1 queue at equilibrium. J. Op. Res. Soc. 1989; 39(2): 183–200.

[9] Kouvatsos, D.D. and Tabet-Aouel, N. A maximum entropy priority approximation for a stable G/G/1 queue. Acta Info. 1989; 27: 247–286.

[10] Tabet-Aouel, N. General queueing networks with priorities. PhD thesis. University of Bradford, Bradford, 1989.

[11] Kouvatsos, D.D. and Almond, J. Maximum entropy two-station cyclic queues with multiple general servers. Acta Info. 1988; 26: 241-267.

[12] Veran, M. and Potier, D. QNAP-2: A portable environment for queueing network modelling. In Potier, D. (ed.) Modelling Techniques and Tools for Performance Analysis. North-Holland, 1985, pp.25–63.

[13] Kouvatsos, D.D. Maximum entropy models for general queueing networks. In Potier, D. (ed.) Modelling Techniques and Tools for Performance Analysis. North-Holland, 1985, pp.589–609.

The CLOWN Network Simulator

Soren-Aksel Sorensen
Mark G. W. Jones

Department of Computer Science
University College London
London WC1E 6BT
United Kingdom

Abstract

The Concatenated LOcal-area and Wide-area Network simulator is a highly modular simulation environment. In its current form, it has been developed to accelerate the performance analysis of a wide range of communication problems. Its window based user interface and macro building facilities simplifies the model building process, and its interactive facilities makes it very suitable for the study of finite horizon problems.

1   Introduction

Performance Analysis is essentially a service oriented subject. Although some investigations may concern themselves with methodology, the end product usually takes the form of advice to a client or another third party. Like most other service oriented subjects, Performance Analysis is constantly faced with two problems: time and presentation. Performance considerations are normally given a very low priority in the Research and Development cycle, and are usually considered only after performance problems are discovered in a product. At that stage in the development cycle, time is a very precious commodity and a heuristic approach is often preferred to a more thorough performance study because the latter is perceived to involve too much time and effort. To shed this image, performance analysts must be able very quickly to provide answers which are of sufficient quality to sustain the client's attention, even if they do not provide the ultimate answer, and present them in a form which is immediately understandable to the client.

2   Requirements

The Concatenated LOcal-area and Wide-area Network (CLOWN) simulator originates from simulation experiments performed in the early eighties [1,2] to support network experiments at University College London [3,4]. Simulation techniques were here used as an investigative tool to support protocol developments. The administrative framework needed for large projects makes it necessary to prepare detailed plans for any experiments well in advance, and the protocol developers loose the ability to perform small investigative experiments of the "what if" type. The best way to provide these facilities is through a simulated environment which closely models the system under development. This type of simulation analysis involves several criteria which are not found in traditional performance analysis. The three most prominent are:

FLEXIBILITY : The simulation model must keep pace with the system developments. Instead of a single model, the investigation involves a series of steps, each a self-contained model representing the system at a stage in its development. These stages must be ready as developers request them. If the simulator falls behind it seizes being useful to the application developers. The model must therefore be very flexible so that changes can be made at a rapid pace. At the same time it is necessary to make sure that validation and verification procedures, which are extremely time consuming, are kept to a minimum.

REALITY : Although developers are interested in the normal performance indices like mean delay and throughput, much of their effort concentrated around fixed limits on delay and resource usage which cannot be exceeded. A majority of the experiments will be concentrated around finite horizon models. Consequently, the models must be able to mimic the detailed behaviour of the system under investigation.

ACCESSIBILITY : Ideally, the developers should themselves be able to build experiments and interpret the data. The building process must therefore be simple and contain elements which are familiar and meaningful to the application developer. The free parameters available to the user must likewise be similar to those provided by the application. The first version of CLOWN [5] was a simple tool-box which assisted the building process by providing a simulation pseudo language through a set of C-callable high level functions. It significantly reduced the time needed to develop simulation models and has been used extensively (sometimes in a form known as the Satellite and LOcal Area Network Emulator) by research students who were interested in the performance aspects of their work but were reluctant to commit themselves to the manpower investment necessary to build simulation models from scratch [6]. A large population of potential users was, however, still excluded either because the network functions provided with the simulator were too restrictive or because the concepts on which the user interface was based were unfamiliar.

3   Structure of CLOWN

The current version of CLOWN has attempted to satisfy the three criteria listed above by adopting a more subject oriented approach. The user is now faced with a library of self-contained network objects which can be combined to form realistic network models. As an example, let us consider the simple network in Figure 1 consisting of five hosts forming a store/forward network.

The same network is shown in Figure 2 translated into the CLOWN semantic. The model is interpreted as a Markov chain with partitioned state table, i.e., each

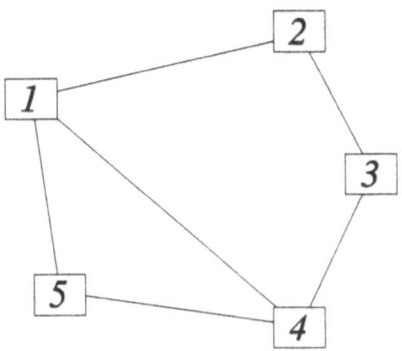

Figure 1: Sample network.

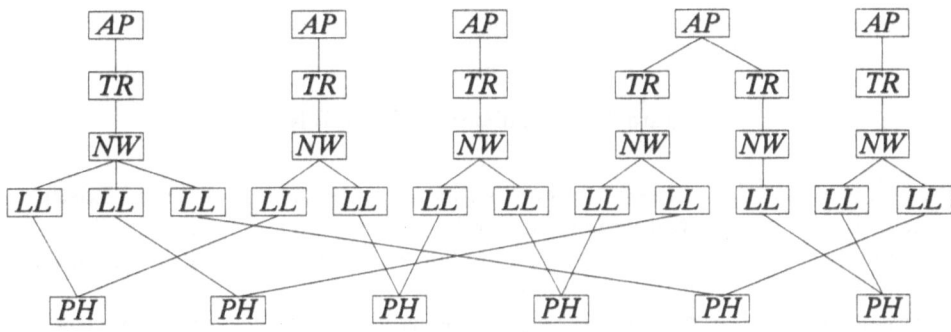

Figure 2: Network graph used by CLOWN.

state transition operates on a specific segment of the state table (a module). The stochastic processes are associated with events which provide the only means by which the modules can interchange information. This approach provides a very general framework suitable for all systems where the information flows at a finite speed. The model in Figure 2 forms a graph constructed from thirty-five modules belonging to five classes shown here as an application layer (AP), a transport layer (TR), a network layer (NW), a link layer (LL) and a physical layer (PH). The links in the graph do not here represent communication links as they did in Figure 1. Instead, they indicate paths along which events can to be transferred. The physical communication links have been replaced by the PH modules. While this example uses generic names for the various classes, CLOWN will incorporate very specific network modules like Ethernets or token rings and protocols like the Transmission Control Protocol (TCP), the User Datagram Protocol (UDP), etc.

A simple block diagram of CLOWN is shown in Figure 3. It consists of two independent processes: the user interface and the runtime process. This gives the user the option to export the runtime process to a more powerful server. Inter-process communication is provided by a proprietary protocol through a standard stream-socket. The simulation is performed by the runtime process. The model is constructed by the experimentation manager using modules from the associated database. Each module is completely self-contained and can be added or deleted independent of the other modules. This provides a substantial degree of flexibility

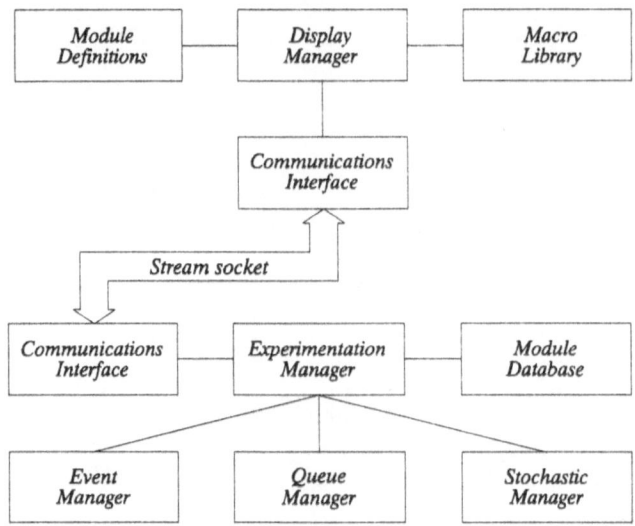

Figure 3: Block diagram of CLOWN.

because new modules can be introduced without any interference with the existing system. If we wanted to test various link level protocols with our example in Figure 2 we could do this simply by adding a new module e.g. LL2 to the database and use it to construct a model where it replaces the LL module. Because no changes we made to any of the existing modules, it would allow us to test the new protocol without interfering with experiments using the older LL module. Several common activities, notably the maintenance of queues and the evaluation of stochastic processes, are not performed by individual modules but by special support processes associated with the experimentation manager.

The simulation model is generated from a description provided by the user interface. A snapshot of the workstation screen is show in Figure 4. It is divided into four areas. The area along the top is a control panel which provides a changing selection of buttons, menus and other control functions. Below the control panel is an assembly surface flanked by two library windows. The left window contains a selection of icons representing the modules available to the user. Copies of these are selected from the library using point and click semantics and placed on the assembly surface where they are connected into a graph. Each icon on the assembly surface represents a separate entity of its type with its own set of parameters. These parameters are manipulated by the user through pop-up windows associated with the icons.

Although the model shown in Figure 2 is fairly simple, the building process can becomes very laborious and repetitive. As an example, we see that the combination AP-TR-NW is repeated four times while the "host" AP-TR-NW-(LL, LL) appears in three places. More complex networks can have even larger structures repeated several times. Apart from the time needed to construct multiple copies of these structures, the repetitive process is prone to errors, and CLOWN compensates for this by allowing the user to define and use macro definitions. A macro is the basic building block used by the CLOWN interface. It contains display information,

127

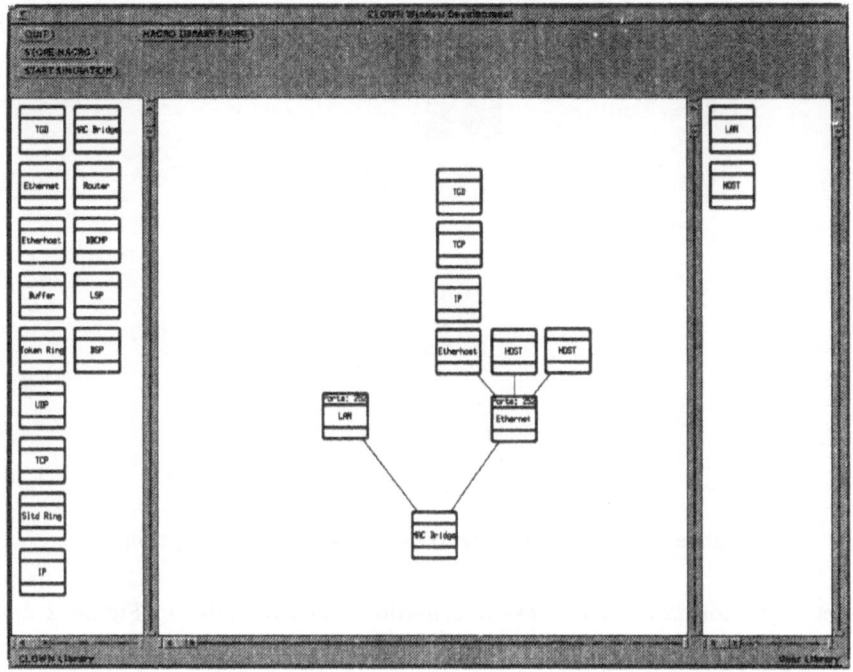

Figure 4: CLOWN user interface during the model building phase.

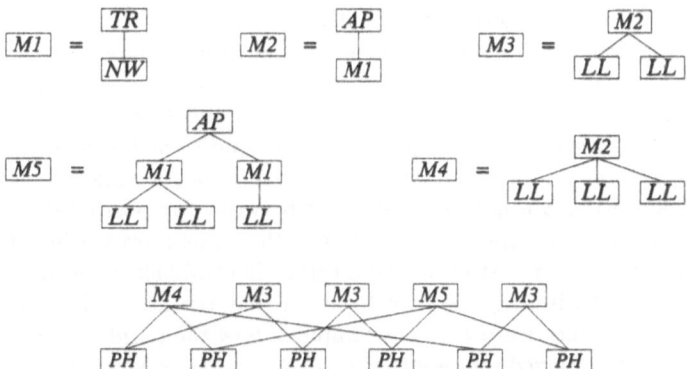

Figure 5: Network model constructed from macros.

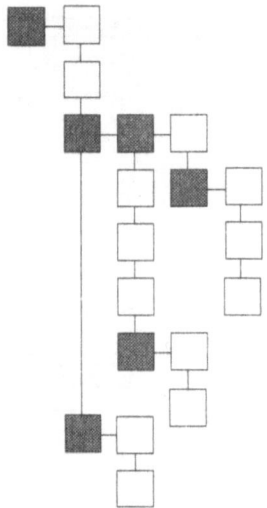

Figure 6: Example of macro usage in model description.

connectivity information and class information. The modules in Figure 2 are in fact macro structures known as "basic" because they contain only one module each. Structured or "composite" macros are collections of basic macros or other composite macros. They also contain the connectivity information for its members. The building process is shown in Figure 4 where the network model has been constructed using macro definitions. The figure shows how composite macros are created and used in the creation of other composite macros to form a hierarchical graph. Macro definitions are stored in a separate library owned by individual users. The macro library is a normal file loaded by the user in contrast to the module library which is intrinsic to CLOWN. Users are thus able to create macro libraries for specific tasks. The structure of model descriptions is outlined in Figure 6 where the filled squares represent composite macros and the open squares basic macros.

A connection between two macros is made by clicking on the appropriate icons in turn. A link is established only if each of the modules has a free connector and these connectors are of matching type. If these conditions are met the connection information is stored and a line drawn between the respective macros, else a warning is issued. Connection can not be established between composite macros. The user must first expand the composite macros until the icons representing the appropriate basic macros appear. After the connection is established the macro can then be collapsed back to its original icon to avoid that the assembly surface becomes cluttered. Figure 5 shows a practical example of how the building process operates. The model being displayed represents two LAN's connected by a MAC-bridge. One of the LAN macros has been expanded to show that it consists of an Ethernet and three hosts. A host has in turn been expanded to show that it consists of a traffic generating application module above TCP/IP and connected through an Ethernet card. The two macros used in the construction can be seen in the macro library window on the right hand side of the assembly surface.

The Ethernet modules in Figure 5 can support 256 potential hosts. Only three have been connected here, but to connect the maximum number of hosts would a daunting task. To ease the process CLOWN has a facility which enables the user to make a 1 : N connection by a single assignment by allowing a single macro to represent N copies. All these copies are identical and must be connected to the same module. The resulting

The runtime process is halted during the construction phase. When this phase is over the process is restarted and the model definition is transferred to the runtime process where the experimentation manager proceeds to build the simulation model after first having performed a consistency check on the received information. During the simulation the user interface provides facilities enabling the user to interact with the model. User can stop, start or pause the process or choose to proceed in short user determined bursts. If the runtime process is inactive the user has the opportunity to change certain of the parameter settings and select how the results are presented. Each module has the ability to maintain its own historical information relating to message rates, queue sizes and other relevant indices. Through pop-up menus associated with the icons on the assembly surface the user is able to select indices from specific modules and display them graphically in separate windows. This allows users to monitor dynamical changes in any part of the simulation. For a more permanent record a hardcopy of the graphical information can be produced at any time.

4   Conclusion

CLOWN was developed foremost to increase the speed with which realistic network models can be investigated. In that capacity it has already proved very useful. We have found that it graphics capabilities make it especially convenient in the study of transient network phenomena because it allows you to monitor several indices simultaneous. One of its applications was to assists the development of a proprietary protocol for a commercial installation. CLOWN was here used during the entire development phase from the initial experiments through validation and verification to the testing of the final product over a simulated network and the customer product demonstration.

The environment has previously been used mainly for research and development activities, but its potential as an educational tool has not been overlooked, and we hope to develop it further in that direction.

References

[1] S-A. Sorensen, "Satellite and Computer Communications", ed. J-L. Grange, pp 343, North-Holland, 1983.

[2] S-A. Sorensen. Cambridge Ring Performance. *Computer Networks and ISDN Systems*, 9(5):345–352, 1985.

[3] P.T. Kirstein et al., "Proceedings of ICCC'82", North-Holland, 1982.

[4] P.T. Kirstein, C.H.C. Leung and S-A. Sorensen, Technical Report 109, Department of Computer Science, UCL, 1984.

[5] S-A. Sorensen, in Proceedings of the 1984 UKSC Conference on Computer Simulation, ed. D. J. Murray-Smith, Butterworths, 1984.

[6] D.Z. Deniz, "Dynamic Channel Management in Packet Mode Usage of Circuit Switched ISDN", Ph.D. Thesis, University College London, 1991.

Modelling ATM Network Components with the Process Interaction Tool

Nick Xenios

Peter Hughes

BNR Europe Ltd.
London Road, Harlow
Essex CM17 9NA
England, UK.

Abstract

This paper reports on a trial application of the Process Interaction modelling tool, PIT, to a class of problems in the area of broadband communication networks. The activity forms part of the Esprit-II project 2143 to develop an Integrated Modelling Support Environment. The specific application was provided by the ATMOSPHERIC project of the RACE programme and is focussed on the modelling of two components: a service access switch and a link multiplexor. In each case the structural details and functionality were kept to a minimum, and attention was directed to modelling the congestion effects of mixed traffic types.

1 Introduction

It is widely accepted that the ISDNs of the next decade will bear services that simultaneously stipulate highly fluctuating transmission rates without compromising the grade of service. Although such services cannot be fully predicted as yet, nevertheless, some have already been identified, as for example, the transmission of high quality video; transmission of "bursty" data resulting from rapid transactions between super-fast computers; and multi-media real-time conferencing.

To support such diverse services the CCITT has adopted Asynchronous Transmission Mode, ATM, as the transport mode of broadband ISDNs. Under ATM, information is transmitted in small packets, or cells as they are now known, in "virtual" channels at rates determined by the service (few Kbps to hundreds of Mbps) rather than by a clock reference within the network. A cell has a size of 53 octets (1 octet = 8 bits) and comprises of a data field (48 octets) and a header field (5 octets); the header field contains information used in virtual path channel identification, multicasting and flow control. Networks with ATM differ from the present ISDNs which are designed to switch information through narrowband channels of fixed capacity (usually 64Kbps), and maintain the channel capacity throughout the duration of the service connection. This type of transmission is referred to as Synchronous Transmission Mode, STM.

From the above discussion it transpires that under STM mode each service tailors itself to the transmission medium, whereas under ATM mode it is the transmission medium that tailors itself to the requirement of each individual service. In other words, an ATM network is bandwidth transparent. The benefits of this characteristic can be summarized as:

- ability to handle variable and varying service patterns;

- ability to respond to new services.

The fact that most existing networks operate under STM, has led to the conclusion that a gradual evolution into full ATM is to be expected. For this reason, collaborative projects like the RACE ATMOSPHERIC [1] have set out to develop hardware platforms that will allow both STM and ATM technologies to be used and interfaced within the same network.

This paper reports on a trial application of the Process Interaction Tool, PIT, to a class of problems in the area of ATM technology. This is part of a set of ongoing activities designed to establish the capabilities of PIT and other elements of the Integrated Modelling Support Environment, IMSE [2], across a wide range of modelling applications. The specific application was provided by the RACE ATMOSPHERIC project in which BNR is a participant, and concerns the two important components of an ATM network: a service access switch and a link multiplexor.

The paper is structured as follows. First, a description of the components to be modelled is given together with the traffic patterns and the quantitative parameters of interest to the designer. This is followed by the adopted modelling approach and the assumptions made with regard to the stochastic behaviour. The results of some modelling experiments with certain design variables are presented, and finally, some inferences are drawn regarding the current and potential benefits of PIT and IMSE with respect to this type of application.

2 Description of the basic ATM components

2.1 The service access switch

The basic function of an ATM service access switch is to route ATM cells across the network on the basis of the information held within the cell headers. Since under ATM there is one class of cells, irrespective of the cell information content, only one type of switch is used.

Each ATM switch is structured around m input and n output queues (ports) which are interconnected by the switch cross-connect fabric (Figure 1). At each input queue the cells are processed in a FCFS (first-come-first-served) fashion, in order to extract the cell-header information and determine the cell destination. Upon the end of the processing delay a cell joins an output port on the switch. The selection of the output port is uniquely determined by the cell next destination-switch along the ATM network; there is a one-to-one correspondence between the output queues of a switch and the switches of the rest of the network. Due to the unscheduled arrival of cells, unlike under STM, there may be contention for the same port, which

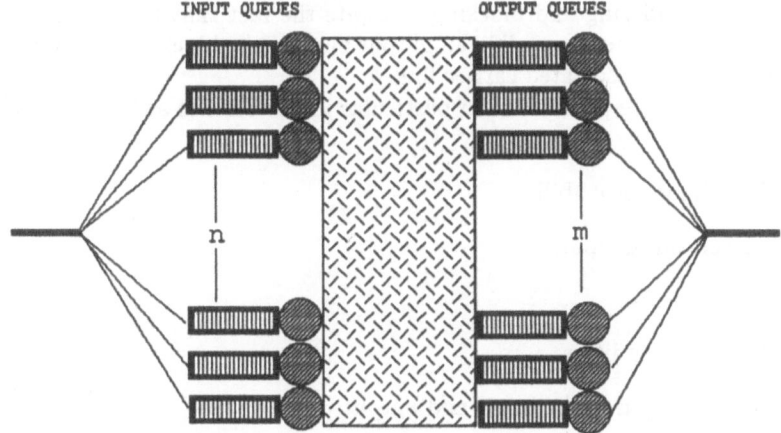

Figure 1: The ATM service access switch.

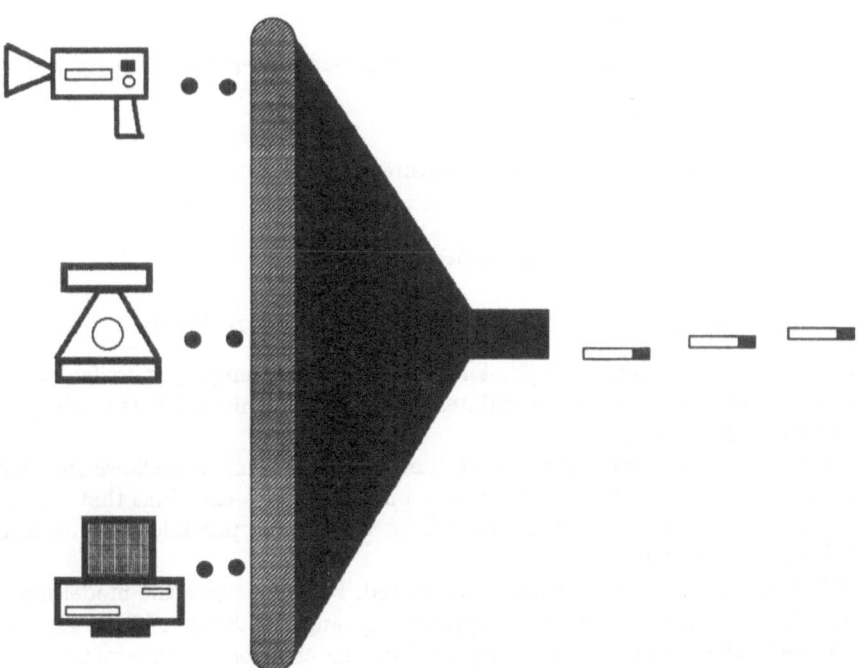

Figure 2: The ATM link multiplexor.

is resolved by buffering and blocking. Despite the fact that there is no processing delay at the output ports, a cell in an output port can be blocked by lack of adequate bandwidth on the associated carrier link.

The performance parameters of particular importance to switch fabric designers are:

1. The cell loss probability.

2. The cell throughput.

3. The cell delay.

4. The cell delay jitter.

5. The utilization of the input and output ports.

2.2 The link multiplexor

The ATM link multiplexor (Figure 2) is a transmission facility consisting of a number of attached input trunks and a FCFS buffer of finite capacity. The main functions of the multiplexor are cell assembly/disassembly, cell statistical multiplexing and bandwidth management. Of interest are the following performance measures for various buffer sizes and input mixes:

1. The statistics of the time periods that the buffer is empty, non-empty and full.

2. The overflow probability.

3. The statistics of the cell interdeparture-times.

3 The modelling approach

3.1 Modelling with the Process Interaction Tool

The Process Interaction Tool [3], known as PIT, was employed to facilitate the process of generating code that simulates the behaviour of models for the components described in the previous section.

PIT is a graphical tool (running on Sun[†][0] workstations) that allows one to draw so called "activity diagrams" [4], which vconsist of nodes and links that depict the life-history for each entity of the model, as well as the possible synchronizations between different entities.

When the model construction is completed, PIT translates the model into DE-MOS, [5] - an extension of the programming language Simula [6]. PIT reduces significantly the amount of time required for the creation of syntactically correct programs, and thus expedites the task of performing simulation experiments with DEMOS. Although PIT does not require the user to master Simula, however, some knowledge of the language is essential. One purpose of this paper study is to show which particular knowledge is required; this is discussed in section 3.3.

---

[0][†] Sun is a trademark of Sun Microsystems, Inc.

Figures 3 and 4 demonstrate the PIT representation of an ATM switch and that of an ATM link multiplexor, respectively. The details of Figures 3 and 4 are explained in sections 3.3 and 3.4 below. For presentation purposes part of the DEMOS code generated for the switch model is given in Appendix C.

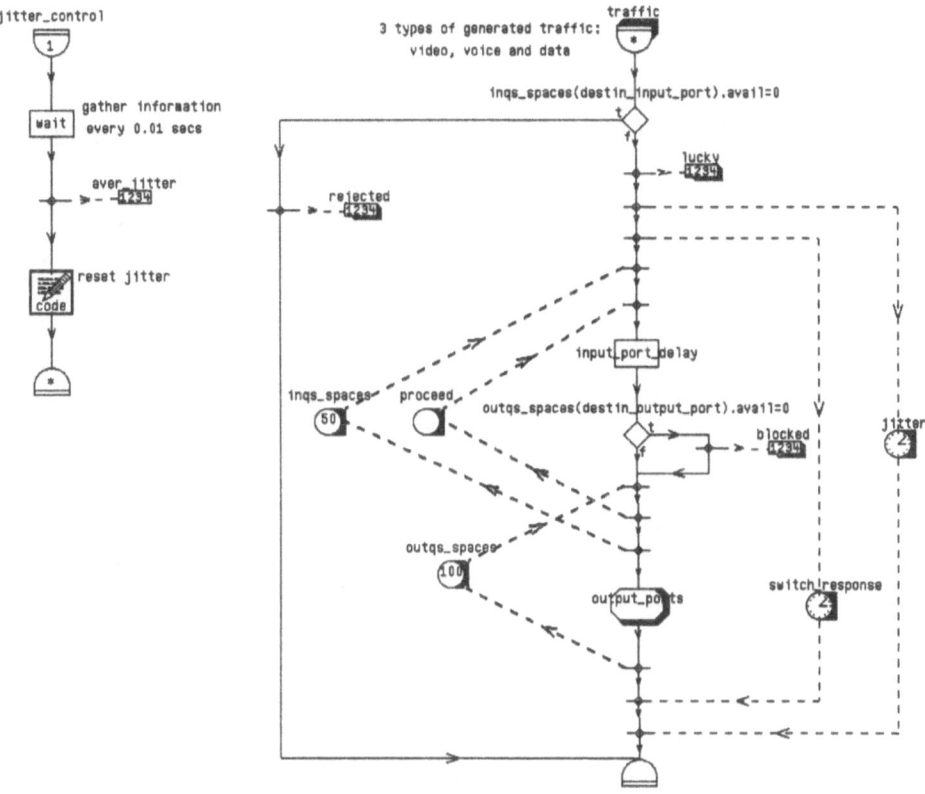

Figure 3: PIT representation of an ATM switch.

3.2 Traffic patterns

For the current modelling study it was assumed that every ATM switch (model 1) receives a traffic mix comprising of 3 streams:

- video at a rate of 34Mbps ($\approx$ 80189cells/s), namely stream S(1);

- voice at a rate of 32Kbps ($\approx$ 76cells/s), namely S(2);.

- and, bursty data at rates in the range 5-100Kbps ($\approx$ 12-236cells/s), namely S(3);

whereas the multiplexor (model 2) receives bundles of traffic consisting of 2 S(1) streams, 100 S(2) streams and 50 S(3) streams.

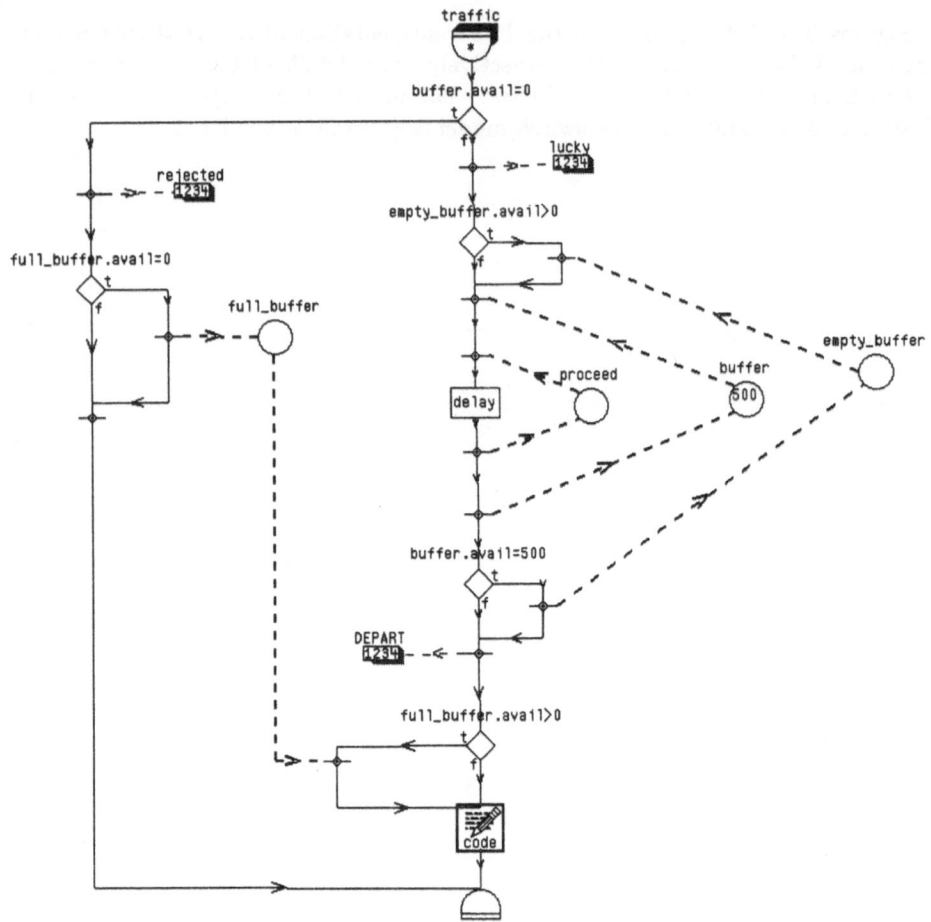

Figure 4: PIT representation of an ATM multiplexor.

A number of assumptions are explicitly made for models 1 and 2 with regard to, in particular, the stochastic delays. To be more specific, the cell interarrival times for streams S(1) and S(2) are taken to follow a negative-exponential distribution with mean 80189^{-1} and 76^{-1}, respectively. For the bursty S(3)-stream the interarrival times between the bursts follow a uniform distributional pattern with lower bound 236^{-1} and upper bound 12^{-1}. Moreover, the size of the bursts is assumed to be geometrically distributed with parameter 0.04.

In model 2 where there are bundles of streams from every stream type, all streams of the same type are merged into one stream by using arguments of stochastic superposition. With this approximation, the DEMOS model is made to consist of 3 streams instead of 152, thus relieving the run-time execution of the simulation from an enormous overhead.

In addition, for each input and output queue of model 1, and for the queue of model 2, the processing delay of every cell is fixed($\equiv$ constant) and its length is a user-given value. Although as mentioned in section 2.1 the output queues of model

1 do not process the cells, nevertheless, the fictitious processing delay at each output queue is interpreted as encapsulating any possible blocking delays.

Finally, for the routing of the cells in model 1 the following routing rules are adopted in this exercise. A cell on its arrival at the switch is routed to the port with the shortest number of queueing cells; however, if the length of the selected input queue equals to the capacity of the queue, then the cell is refused entry, and is thereupon lost. In contrast, the selection of an output queue is randomly determined.

3.3 The switch model

The model of Figure 3 is explained in terms of PIT and DEMOS concepts introduced in *italics*. For further explanation the reader is referred to [3]-[5].

In the switch model cells are queued in input and output buffers which are represented by arrays of *resource* objects "inqs_spaces" and "outqs_spaces". Resource objects in DEMOS have their own internal queues which the modeller does not have to manage explicitly. Processing time within the switch is represented by a simple *hold* box, "input_port_delay". Single thread processing at the server is assured by the "proceed" resource. Processing at the output buffers is represented by a set of servers, one for each buffer. A *server* object in PIT is a convenient shorthand for the common sequence depicted below.

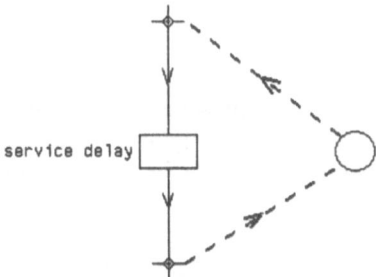

Traffic flow is represented by the life-history of an *entity* of class "traffic". This is an arrayed class of dimension three, corresponding to the main traffic types. Each instance of a traffic entity represents an ATM cell.

The statistical behaviour of the traffic is obtained by instrumenting the traffic entity appropriately. Thus there are *counters* for "lucky" and "rejected" where the latter represents cell loss due to full input buffers. There is a counter for cells which are temporarily blocked due to full output buffers. Switch response time is computed by means of a *clock* object. This maintains statistics on the elapsed time between two clocking points, including mean, standard deviation, min, and max.

So far we have described standard facilities offered by PIT, which do not require any special knowledge. It is interesting to note which additional features were found to be useful for this particular model.

inspection of resource queues

It is frequently useful to be able to inspect the state of a resource object before deciding on a course of action. This is done by accessing the standard attribute *avail* using the Simula "dot-notation". This technique was used to monitor blocking at the input and output ports.

selection of input/output queues

As noted in the previous subsection, cells are allocated to input and output queues according to specific rules. These rules are implemented as follows. The global procedure *short_queue* uses the standard attribute *avail* of DEMOS resource objects to find the index "destin_input_port" of the input queue with the shortest length. The output queue index "destin_output_port" is selected by drawing a random integer in the appropriate range.

computation of jitter

Jitter is defined as the fluctuation of switch response time within a fixed time-interval. It is computed by inserting an extra *clock* object "jitter", which is reset periodically. An auxiliary entity *jitter-control* was introduced at the PIT level, which resets the jitter clock every .01 seconds after recording the fluctuation "jitter.max - jitter.min" using a *code*-module. A *counter* object "aver_jitter" is used to maintain statistics on the fluctuation.

All model specific information is provided by editing the "Model attributes" form (Figure 5). This has entries for specifying the trace, simulation and report control options, as well as the declarations of the model variables, procedures and distributions.

3.4 The multiplexor model

As in the switch model, cells are represented in model 2 by the instances of the arrayed entity "traffic". While in the multiplexor, all cells experience in FCFS order a delay which has a user-given value. This delay is depicted in Figure 4 by the *hold* box "delay".

The maximum capacity of the multiplexor is specified by the number of tokens of the "buffer" *resource*. The availability of the "buffer" tokens determines the acceptance or rejection of incoming cells. Moreover, the "empty_buffer" and "full_buffer" *resources* facilitate the gathering of the statistics for the times the multiplexor is in empty and full states; the utilizations of "empty_buffer" and "full_buffer" correspond to the percentage of time the multiplexor is empty and at full capacity, respectively.

Finally, the statistics of the cell interdeparture-times are monitored by the "DEPART" *counter* which uses the time of the last cell departure as marked by the Simula code included in a *code* box.

4 Simulation results

A representative but by no means exhaustive simulation study was conducted in order to obtain estimates of the performance measures of models 1 and 2 on a Sun-4 workstation capable of 16Mips. For each simulation experiment 6-8 replications were needed and each replication was run for 0.3 simulation seconds to assure steady state results.

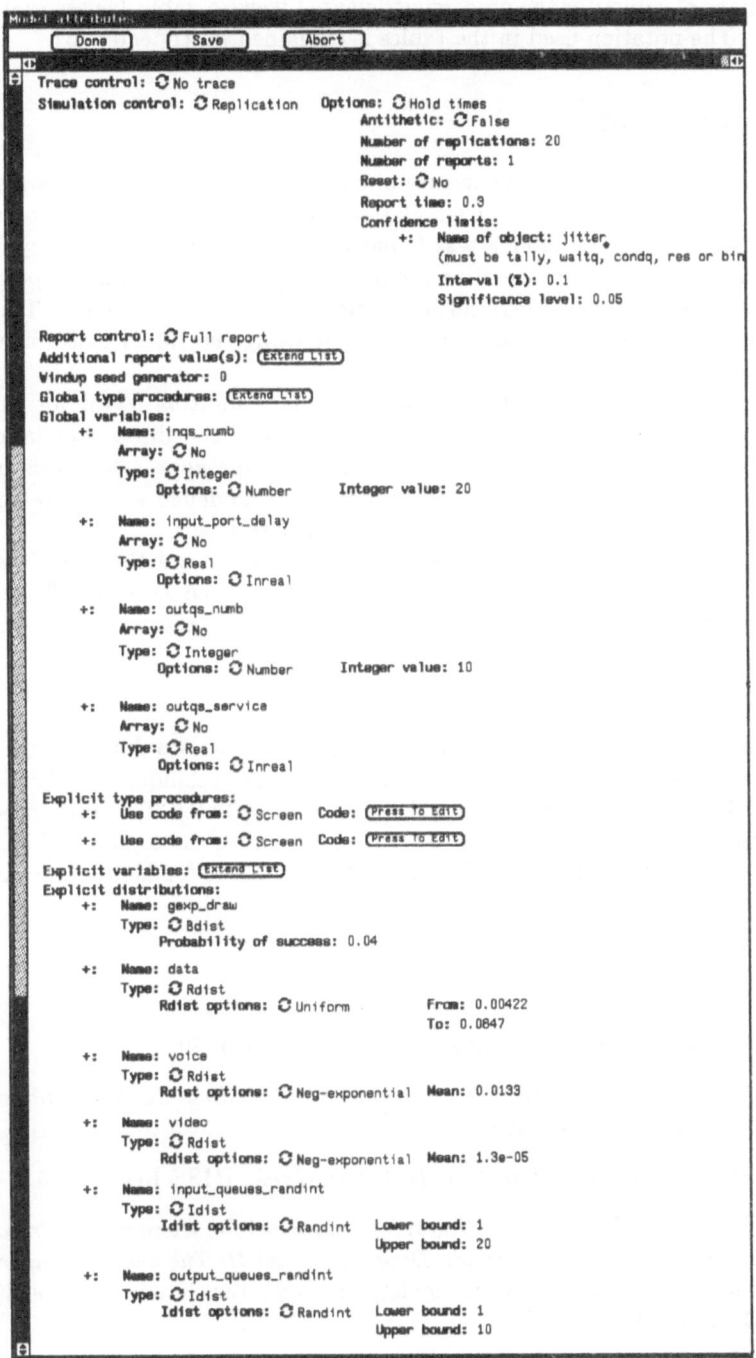

Figure 5: Attributes form of model 1.

Table 1 illustrates the effect on the performance measures of model 1 of various combinations of values for the processing delays. Likewise, table 2 is associated with model 2. (The notation used in the tables is explained in Appendix B).

5   Conclusions

This exercise demonstrates the capacity of PIT as a tool for representing models in a graphical way and for producing simulation code associated with the models. The time for constructing the models was found to be negligible (one man-day for each model) when compared to the time required for setting experiments, performing simulation runs, and collecting and organizing the simulation results. This experience underlines the benefit to be obtained by using the IMSE platform when it becomes available.

Generally, the simulation time requirement is a limiting factor for exploring a wide range of design cases. This clearly rules out the direct simulation of a network having a realistic number of components. One approach to this problem is hierarchical modelling, in which components are modelled separately and then recombined in a high-level model with some loss of accuracy.

More recent versions of PIT support this technique, and IMSE will assist its application through the use of hierarchical experiments. (A possible PIT representation of an ATM network using an array of submodels is shown in Figure 6).

Acknowledgements

The modelling requirements used for the two components are due to Dr M. Anagnostou (National Technical University of Athens). The technique used for the computation of the jitter was devised by Dr E. Barber (BNR). The authors are also grateful to M.J. Hillyard (BNR) for introducing them to the topic of ATM, and to Dr S. Bruyn (BNR), ATMOSPHERIC project manager, for his part in initiating this exercise.

References

[1] RACE ATMOSPHERIC (R1014). *Annual Report*. 1989.

[2] P.H.Hughes and D.Potier. *The Integrated Modelling Support Environment*. IMSE Report R-1.2-4, 1989.

[3] E.O.Barber. *Process Interaction Tool User Guide*. IMSE Report R-5.1-4, 1991.

[4] R.J.Pooley and P.H.Hughes. *Towards a Standard for Hierarchical Process Oriented Discrete Event Simulation Diagrams. Part II: The suggested approach to flat models*. Transactions of the Society for Computer Simulation, Vol. 8(1), pp. 21-31, March 1991.

[5] G.M.Birtwistle. *Discrete Event Modelling on Simula*. Macmillan, 1985.

[6] R.J.Pooley. *An Introduction to Programming in SIMULA*. Blackwell Scientific, 1987.

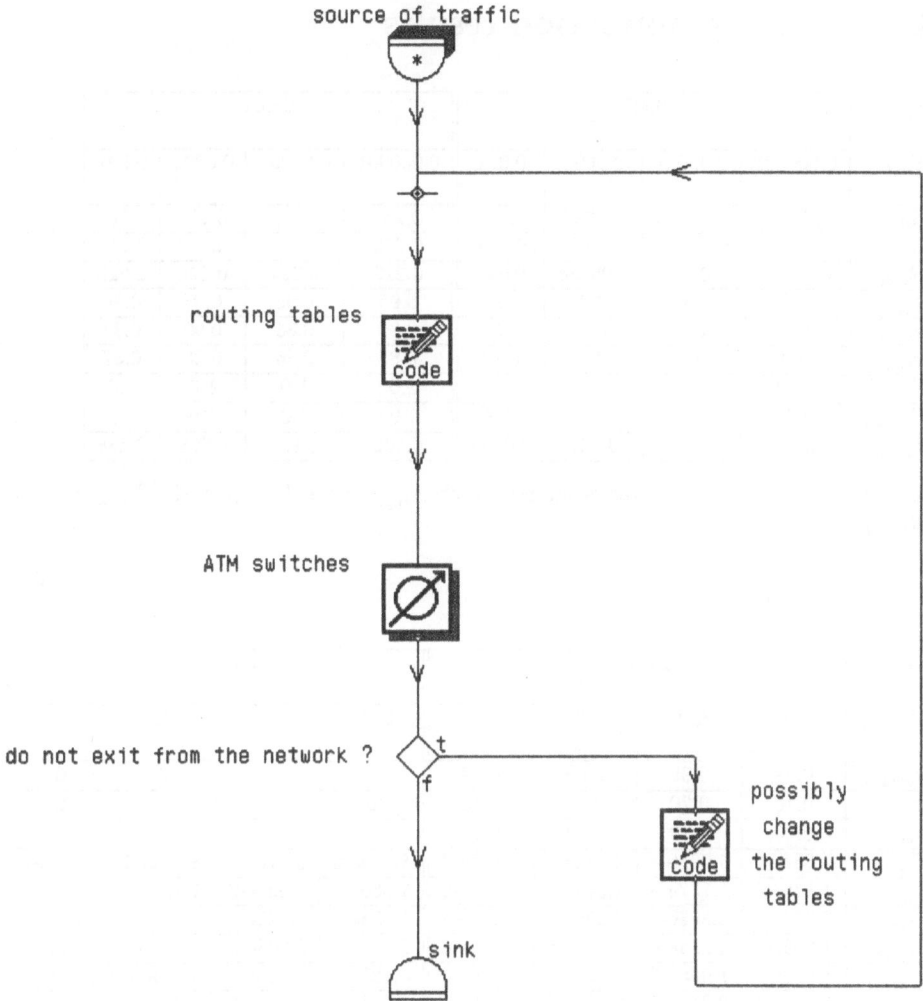

Figure 6: PIT representation of a network of ATM switches.

A   Tables of simulation results

S_{output}	0.0001				0.001			
S_{input}	0.000001	0.00001	0.0001	0.001	0.000001	0.00001	0.0001	0.001
R_{data}	0.00052	0.00053	0.00062	0.082	0.167	0.142	0.109	0.160
R_{voice}	0.00028	0.00029	0.00038	0.097	0.205	0.230	0.218	0.212
R_{video}	0.00028	0.00029	0.00038	0.098	0.222	0.224	0.222	0.223
π_{data}	0.0	0.0	0.0	0.92	0.95	0.96	0.97	0.96
π_{voice}	0.0	0.0	0.0	0.62	0.86	0.85	0.90	0.86
π_{video}	0.0	0.0	0.0	0.73	0.85	0.86	0.86	0.87
U_{input}	0.00	0.00	0.00	1.0	1.00	1.0	1.0	1.0
U_{output}	0.77	0.78	0.78	0.2	1.00	1.0	1.0	1.0
J_{video}	0.001	0.001	0.001	0.0008	0.202	0.182	0.208	0.138

Table 1: Performance measures of model 1 with $N_{input} = 20$, $N_{output} = 10$, $C_{input} = 100$, $C_{output} = 50$

S_{buffer}	0.00001				0.000001			
C_{buffer}	2000	3000	4000	5000	2000	3000	4000	5000
π_{data}	0.99	0.99	0.99	0.99	0	0	0	0
π_{voice}	0.70	0.70	0.70	0.70	0	0	0	0
π_{video}	0.70	0.70	0.70	0.70	0	0	0	0
D_{data}	7.354e-4	6.601e-4	5.842e-4	5.163e-4	8.176e-6	8.176e-6	8.176e-6	8.176e-6
σ_{data}	1.175e-4	1.137e-4	1.043e-4	7.082e-5	1.850e-6	1.850e-6	1.850e-6	1.850e-6
D_{voice}	2.249e-4	2.253e-4	2.258e-4	2.262e-4	1.223e-4	1.223e-4	1.233e-4	1.223e-4
σ_{voice}	1.225e-6	1.214e-6	1.237e-6	1.186e-6	2.708e-5	2.708e-5	2.708e-5	2.708e-5
D_{video}	1.063e-5	1.065e-5	1.067e-5	1.070e-5	5.723e-6	5.723e-6	5.723e-6	5.723e-6
σ_{video}	1.763e-8	2.210e-8	2.693e-8	2.433e-8	1.278e-6	1.278e-6	1.278e-6	1.278e-6
U_{buffer}	0.99	0.99	0.99	0.99	0.20	0.15	0.09	0.003
F_{buffer}	0.22	0.22	0.22	0.22	0	0	0	0

Table 2: Performance measures of model 2 with $S_{buffer} = 0.00001, 0.000001$

B Notation of tables 1-2

$N_{input/output}$	Number of each input/output queue in model 1.
$C_{input/output}$	Capacity of each input/output queue in model 1.
C_{buffer}	Capacity of the buffer in model 2.
$S_{input/output}$	Processing delay at each input/output queue in model 1.
S_{buffer}	Processing delay in the queue of model 2.
$U_{input/output}$	Utilization of each input/output queue in model 1.
U_{buffer}	Utilization of the buffer in model 2.
F_{buffer}	Proportion of time the buffer of model 2 is full.
$\pi_{data/voice/video}$	Overflow probability of data/voice/video cells.
$R_{data/voice/video}$	Average delay of data/voice/video cells in model 1.
$D_{data/voice/video}$	Average interdeparture-times of the data/voice/video cells.
$\sigma_{data/voice/video}$	Interdeparture-times' standard deviation of the data/voice/video cells.
J_{video}	Delay jitter of the video cells.

C DEMOS code for model 1

```
begin
    !! *** DECLARATION OF GLOBAL VARIABLES *** ;
    integer klm, inqs_numb, outqs_numb;
    real    input_port_delay, outqs_service;

    !! *** INSTANTIATION OF GLOBAL VARIABLES *** ;
    hold_time:=inreal;
    inqs_numb:=20;
    input_port_delay:=inreal;
    outqs_numb:=10;
    outqs_service:=inreal;

    demos
    begin

    !! << declarations of resources, tallies, counters, etc come here >> ;

    !! *** DECLARATION OF EXPLICIT PROCEDURES *** ;
      real procedure select_traffic(index);
      integer    index;
      begin
        boolean toss;
        real as;
        if index=1 then
```

```
  begin
    toss:=gexp_draw.sample;
    as:=data.sample;

    if toss then select_traffic:= as
            else select_traffic:= 0.0
  end
    else if index=2 then select_traffic:=voice.sample
                    else select_traffic:=video.sample;
end ***select_traffic***;


integer procedure short_queue( wq );
ref(res) array wq;
begin
  integer n,i,j,k;
  k := 1;
  n := wq(1).avail;
  for i:=2 step 1 until inqs_numb do
  begin
    if wq(i).avail > n then
    begin
      k := i;
      n := wq(i).avail;
    end;
  end;
  short_queue := k;
  end ***short_queue***;


!! *** DECLARATION OF EXPLICIT DISTRIBUTIONS *** ;
ref(bdist) gexp_draw;
ref(rdist) data, voice, video;
ref(idist) input_queues_randint, output_queues_randint;


!! ENTITY DEFINITIONS ;
entity class traffic_kicker(index);
integer index;
begin
  integer destin_input_port, destin_output_port;
  loop:
  destin_input_port:=short_queue(inqs_spaces);
  destin_output_port:=output_queues_randint.sample;
  new traffic("traffic",index,destin_input_port,destin_output_port).schedule(0.0);
  hold(select_traffic(index));
  repeat;
end *** traffic_kicker *** ;


entity class traffic(index,destin_input_port,destin_output_port);
integer index, destin_input_port, destin_output_port;
begin
  !! << body of the traffic entity comes here >> ;
end *** traffic *** ;
```

```
entity class jitter_control;
begin
  !! << body of the jitter_control entity comes here >> ;
end *** jitter_control *** ;

!! INITIALISATION ;
gexp_draw          :- new draw("gexp_draw",0.040000);
data               :- new uniform("data distribution",0.004200,0.084700);
voice              :- new negexp("voice distribution",1.0/0.013300);
video              :- new negexp("video distribution",1.0/0.000013);
input_queues_randint:- new randint("input_queues_randint",1,20);
output_queues_randint:- new randint("output_queues_randint",1,10);

!! << initialisation of resources, tallies, counters, etc come here >> ;

!! RUNTIME PARAMETERS;
  while confidence(jitter(1).sdaverage,0.100000,0.050000)
    and confidence(jitter(2).sdaverage,0.100000,0.050000)
    and confidence(jitter(3).sdaverage,0.100000,0.050000)
  and replication <= 20 do
  begin
    for klm:=1 step 1 until 3 do
        new traffic_kicker("traffic_kicker",klm).schedule(0.0);

    new jitter_control("jitter_control").schedule(st);

    hold(0.3);
    if replication > 20 then noreport;
    replicate;
  end;
end;
end;
```

Asynchronous Packet-switched Banyan Networks with Blocking (Extended Abstract)

Peter G. Harrison
Afonso de C. Pinto

Department of Computing
Imperial College
180 Queen's Gate, London SW7 2BZ

Abstract

We derive an approximate algorithm to predict the mean transmission time through finitely buffered, packet-switched, asynchronous networks with no feedback. We tailor the algorithm to banyan networks, which are important in parallel computer architectures and telecommunication systems, and present preliminary numerical results. The full paper appears in [3].

1   Introduction

As computer and communication technology continues to advance, there is an increasing need for efficient communication networks. As traffic increases in asynchronous, packet-switched networks, blocking occurs at the nodes and it is this problem that we address. Our analysis applies to networks with no feedback which can always be structured in stages, some of which can be skipped by a packet. The components which forward messages in each stage are modelled independently as small sub-networks of queues; these components are *crossbar switches* in the case of Banyan networks. The approach uses an iterative algorithm of the type used by Jun and Perros for tandem networks [4] but we generalise to feed-forward networks which need not be tandem. If we consider a particular packet, its path through the network can be viewed as an open tandem network with external arrivals at each node. Essentially, it is the handling of these arrivals, which are not independent of the said packet nor of each other, which constitutes the generalisation.

Although our analysis applies to any feed-forward network, we focus on regular Banyan networks which are especially popular in parallel computer architectures and in the ATM switches prevalent in telecommunication network designs. Various models for multi-stage interconnection-networks have been developed, corresponding to different protocols, modes of buffering and synchronisation, [7]. Many assume synchronous operation, e.g. [5], which generally simplifies the analysis since packets are transmitted on fixed clock pulses. Hence independence between any arrival processes can be considered at each stage at the same discrete time instants without regard for past history; there is never any backlog of work since the network is effectively cleared before each clock pulse. At the other extreme, circuit-switched asynchronous banyans have been studied by Harrison and Patel [2]. In this case, past history plays a crucial role and the system is modelled in the steady state by using Little's result and exploiting the recursive structure of the network extensively. In between these extremes comes packet-switched networks which can be modelled as standard queueing networks if the buffers never overflow. Indeed, in this case, not only can throughput and mean values be determined, but also the *distribution* of transmission time; see [1]. The present analysis therefore considers a more realistic system – buffers are rarely infinite! However, it is restricted to analysis of mean values in view of the hugely increased complexity arising from the absence of a product form solution for queueing networks with blocking.

2   Technical summary

The analysis and the resulting algorithm is based on the following principles. First we define some terms. At each stage of a Banyan, there is a set of inputs and a set of outputs. These are numbered (arbitrarily) from the top of a network. The same applies to the switching components in a stage, i.e. the crossbars, and the sequence number of each component is called its level in that stage.

The first step of the algorithm computes the joint probability distribution of the queue-lengths in those components of each stage–j that can cause each other to become blocked. Blocking may arise because the components' outputs are connected to a common component in the next stage, $j + 1$, to the right. This component may be full due to packets from one such component in stage–j, all components of which are then blocked. The iteration begins at the rightmost (i.e. output) stage which is not blocked and propagates back to the first. Blocking is represented by the state of the common queue in stage–j which is artificially extended from its original capacity, c say, to $c + 2$. Then blocking occurs when the queue length is c (next packet is the first in the blocked queue), $c + 1$ (one packet is already blocked) or $c + 2$ (there is no input to the queue until a service completion occurs). The different scenarios corresponding to these three states are crucial to the analysis. Notice, however, that the final stage is never blocked since we assume that all outgoing packets from the network are either accepted or else lost. This enables the step to commence. Finally, for step 1, all internal arrival processes are assumed Poisson, in the absence of better information.

The second step works from left to right and determines the output process of each component in the current stage in terms of its input process using approximate results on the superposition and splitting of arrival processes, [6], [8]. The input process to the first stage is given — here Poisson. The mapping function from input to output process at each component in a stage depends on the joint distribution of queue lengths determined in the previous step.

The two steps are then repeated until convergence is achieved, measured by the closeness of successive approximations to the joint queue length probability distributions. In the odd-numbered steps, however, the internal arrival processes used are those computed in the previous, even-numbered, step, rather than being assumed Poisson as in step 1.

Each crossbar comprises a single buffer into which arrivals from the two input pins enter. This is modelled by an open queueing system of two servers, each with its own queue, with a given maximum population equal to the size of the buffer. The balance equations of the crossbar sub-model are then solved directly. Service times are taken to be two-stage Coxian, which can often provide a good approximation to general service times. However, an alternative distribution is the Generalised Exponential which also performs well and results in a more efficient model.

Preliminary numerical results of our algorithm are given in [3] and numerical validation of our model with respect to simulation will follow. The results show, through the probability distributions of the number of packets in each crossbar and, in particular, their means, how the hot-spot effect well known in Banyan networks builds up in the decode tree leading to the hot-spot. Moreover, fully general access patterns to the outputs of the network can be handled.

3 Conclusion

We have derived a novel algorithm to predict the performance of asynchronous, packet-switched banyan networks with finite buffers in the crossbars, i.e., with blocking. This is based on a known overall approach which has been used effectively to model blocking in tandem networks of queues. However, it handles more complex dependencies between switches arising from packet routings as well as blocking at downstream crossbars. Preliminary numerical results have been obtained and validation of the approximate algorithm with respect to simulation is underway.

References

[1] P. G. Harrison, "On nonuniform packet switched delta networks and the hot-spot effect", *IEE Proceedings-E*, **138**, no 3, pp. 123–130, May 1991.

[2] P. G. Harrison and N. M. Patel, "The Representation of Multistage Interconnection Networks in Queueing Models of Parallel Systems," *Journal of the ACM*, **37**, no 4, pp. 863–898, Oct. 1990.

[3] P. G. Harrison and A. de C. Pinto, "Blocking in asynchronous, buffered Banyan networks", In proc. *Int. Conf. on the Performance of Distributed Systems and Integrated Communication Networks*, Kyoto, Japan, Sep. 1991.

[4] K. P. Jun and H. G. Perros, "An Approximate Analysis of Tandem Queueing Networks with Blocking and General Service Times," *European Journal of Operational Research*, **46**, pp. 123–135, 1990.

[5] H. S. Kim and A. Leon-Garcia, "Performance of Buffered Banyan Networks under Nonuniform Traffic Patterns," *IEEE Trans. Comm.*, **38**, no 5, pp. 648–658, May 1990.

[6] P. J. Kuehn, "Approximate Analysis of General Queuing Networks by Decomposition," *IEEE Trans. Comm.*, **COM-27**, no 1, pp. 113–126, Jan. 1979.

[7] J. H. Patel, "Performance of Processor-Memory Interconnections for Multiprocessors," *IEEE Trans. Comput.*, **C-30**, pp. 771–780, Oct. 1981.

[8] W. Whitt, "The Queueing Network Analyzer," *The Bell System Technical Journal*, **62**, no 9, pp. 2779–2813, Nov. 1983.

Equilibrium Point Analysis of a Slotted Ring

Michael E. Woodward
Department of Electronic & Electrical Engineering
Loughborough University of Technology
Loughborough, Leics, LE11 3TU
United Kingdom

Abstract

The access protocols for computer communication networks can, in principle, be modelled as multidimensional Markov chains. In most practical systems however, the state-space is so vast that a solution by classical Markov analysis is intractable, and some approximation technique must be used. In this paper the access protocol of a slotted ring network is modelled as a discrete-time Markov chain, and its solution is obtained by equilibrium point analysis (EPA), a method that approximates the stationary probability distribution of the Markov chain by a unit impulse located at a point in the state-space where the system is in equilibrium. The model is found to give good results over a wide range of parameter values when compared with an equivalent simulation study.

1   Introduction

In this last decade of the twentieth century, the so called information revolution has brought about the need to be able to efficiently interconnect and interact with a large number of information processing devices that are situated at geographically diverse locations. From an analytical point of view, perhaps one of the most interesting networks is that known as the slotted ring, which can be used for interconnecting devices within a local area, such as a university campus for example. This system was independently proposed by Farber and Larson [4], and Pierce [10], and has since been developed along more commercial lines as the Cambridge ring (Needham and Herbert [9]) and the Orwell ring (Falconer and Adams [2]).

In its basic form the slotted ring is very simple; rather deceptively so, as we shall see. It consists of a ring of cable, or optical fibre, along which signals propagate in one direction only. Spaced at various intervals around the ring are the network *stations*, the purpose of which is to connect the various information processing devices to the ring, which then acts as a transmission medium between these devices. At any given time there can be perhaps several hundreds of bits in flight, travelling around the ring, which is divided into an integral number of *slots*, where each slot is considered to be sufficiently large to contain a single, fixed length packet. Each of the stations maintains a *buffer* for queuing packets awaiting access to the ring for transmission to some other station.

The algorithm that all stations use for controlling access of their packets to the ring is the *medium access control* (MAC) *protocol*, and for a slotted ring this can function as follows. Slots circulating around the ring are designated as either full or empty, depending on whether or not they contain a packet, respectively, and the first bit of a slot acts as an indicator to denote the full/empty status of the slot. Stations that have packets to transmit simply wait for the next empty slot to pass by and load this with the packet at the head of their queue. The packet, which thus proceeds on its way around the ring, contains the address of a destination station which, when recognised by the destination station, causes that station to copy the packet into its memory. The slot, still marked as full, continues on its way around the ring until it arrives back at the source station, which then marks the slot as empty and passes the slot on to the next downstream station. Any station that holds a slot on the ring is not allowed to transmit a second packet until the slot has been emptied, nor is a station that has just emptied a slot allowed to reuse the same slot. These latter two rules are to ensure that all stations have a fair share of the system's transmission capacity.

In the foregoing it should be noted that we have ignored any error recovery procedures, but usually errors are so infrequent, particularly when optical fibre waveguides are used as the transmission medium, that any performance overheads due to transmission errors can be ignored without measurably affecting the performance results in any way.

A number of previous studies have addressed the determination of the performance of a slotted ring in terms of the throughput and mean packet access delay. Among the more notable are those by Bux [1], Harrus [5], King and Mitrani [7], and Kamal and Hamacher [6]. All previous studies however are characterised by either considering each of the network's stations to have a buffer with space for, at most, one packet (or perhaps one multipacket message), or to consider the stations to have infinite buffer storage. Whilst both viewpoints are unrealistic in a practical context, the latter precludes the determination of packet loss performance, a parameter that is critical in assessing the transmission quality for certain types of data, such as voice or video. The most original feature of our study is that we consider the network's stations to each have a buffer of specified (finite) size, and are thus able to obtain packet loss performance measures, in addition to the usual throughput and mean packet access delay.

2 Model Formulation

In what follows we shall formulate a discrete-time Markovian model of the MAC protocol for the slotted ring that was described in the previous section. The following assumptions will be made.

- The ring has N stations and M slots, where $N \geq 2$ and $M \leq N$.

- Stations behave independently, and have identical statistical parameters.

- Stations are spaced equidistantly around the ring.

- Packets are of fixed length, and fit exactly into a slot.

- Slots behave independently.

- Time is measured in discrete units called *slot times*, where a slot time is $1/M$ of the total time for a slot to circumnavigate the ring.

- The interarrival time (in units of slot times) of packets at a station is geometrically distributed with mean $1/p$. That is a station generates a packet with probability p per slot time, with the packet generation assumed to occur at the start of a slot time.

- Each station has a finite length buffer of maximum capacity J packets, and has a first-in first-out queuing discipline.

- If a packet departs from and a new packet arrives at a stations's buffer in the same slot time, the former is assumed to take place before the latter.

- Each station is allowed to transmit only one packet at a time.

- When a packet gains access to a slot on the ring, the packet is immediately removed from the head of the station's buffer.

- a station can generate, at most, one new packet during that station's transmission period; this packet generation is assumed to occur at the start of the last slot time of a transmission period with probability Mp.

- The transmission medium is error free.

The penultimate assumption is to simplify the modelling, and its implications will be considered in more detail later. Clearly, this assumption restricts the possible values of p to the range $0 \leq p \leq 1/M$.

Under the above assumptions, a model for the MAC protocol of a slotted ring is shown in Fig 1. It should be noted at the outset that Fig 1 is not a state-transition diagram, at least not in the conventional sense, but is a higher level diagram where each of the circles represents a mode (it will later be seen that the modes correspond to the components of a state vector in a multidimensional Markov chain). Each station can be in one of the modes at a given slot time, and can either remain in the same mode or move into another mode at the end of the slot time, mode transitions thus taking place each slot time.

Each mode in Fig 1 is labelled (i,j), $0 \leq i \leq J, 0 \leq j \leq M$, where i denotes that stations in mode (i,j) have i packets in their buffer, and j notes the *transmit status* of stations in mode (i,j). If $j = 0$ then the stations are not transmitting, otherwise for $1 \leq j \leq M$ stations are transmitting on the ring, and will have completed transmission of their packet in $M - j$ slot times after the current one.

Fig 1 also shows the *mode transition probabilities*, and these represent the probabilities of stations in a given mode moving into another mode at the end of a slot time. If S' is the probability that stations in $(i,0)$ modes, $0 \leq i \leq J$ (non-transmitting stations), find the next passing slot full, then the transitions between the various modes are as in Fig 1. For example, stations in mode $(0,0)$ are idle and have no waiting packets, and these stations will thus remain in $(0,0)$ mode at the end of a slot time if they do not generate a new packet (probability $1 - p$). The stations will transmit a packet and move to mode $(0,1)$ at the end of the slot time if

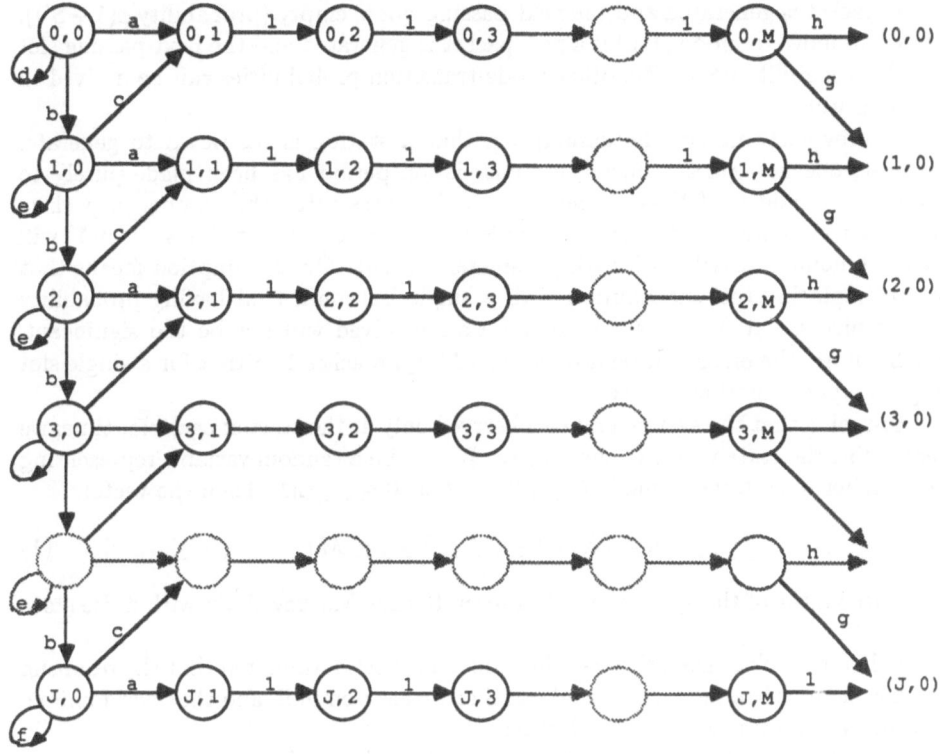

where - a = p(1-S')   d = 1-p   g = Mp
       b = pS'   e = (1-p)S'   h = 1- Mp
       c = (1-p)(1-S')   f = S'

Fig.1   Markov model for a Slotted Ring.

a new packet is generated and the next passing slot is empty (probability $p(1-S')$). They will move to mode $(1,0)$ if a new packet is generated and the next passing slot is full (probability pS'). The other mode transition probabilities can be derived in a similar way.

As previously noted, the assumption that a station is restricted to generate, at most, one new packet during a transmission period has been made purely to simplify the model. If Y is a random variable representing the number of packets generated by a station during a transmission period of M slot times, then Y will have a binomial distribution with parameters (M,p). Our assumption means that we are replacing this distribution with a single Bernoulli trial having probability Mp of success. If p is small, then the error involved will not be too significant. Furthermore, the error will tend to zero as M approaches 1 in that for a single slot ring the approximation is exact.

Assuming that the system state is observed only at the start of each slot time, we next define the state vector of the system. Let n_{ij} be a random variable representing the number of stations in mode (i,j), $0 \leq i \leq J$, $0 \leq j \leq M$. Then the vector

$$\mathbf{n} = (n_{ij} : 0 \leq i \leq j, 0 \leq j \leq M) \tag{1}$$

is a state vector of the system, and is a discrete-time Markov chain with finite state space.

It should be clear from the way the model has been formulated that the resulting Markov chain is irreducible, aperiodic and recurrent non-null, and thus has a unique stationary state probability distribution.

Since there can be up to N stations in each of the modes $(i,0)$, $0 \leq i \leq J$, and at any given time there can be up to M stations transmitting on the ring (one in each slot), then $0 \leq n_{i0} \leq N$, $0 \leq i \leq J$, and $0 \leq n_{ij} \leq M$, $0 \leq i \leq J$, $1 \leq j \leq M$. Although the components of the state vector are strongly dependent on each other, it takes little imagination to realise that, for all but very small values of N, M, and J, the solution of such a multidimensional Markov chain to obtain the stationary state probability distribution would be intractable by classical Markov analysis, and some approximation technique must be employed to find a solution.

3   Solution of the Model

To solve the model we shall use equilibrium point analysis (EPA), which is described in detail in the monograph by Tasaka [11]. In essence, EPA replaces the stationary state probability distribution of a Markov chain by a unit impulse located at a point in the state space where the system is in equilibrium (the equilibrium point). This avoids the calculation of the one-step state transition probabilities of the Markov chain, and the solution reduces to finding the values of the components of the state vector at the equilibrium point.

The mathematical principle of EPA is very simple: Namely that the conditional expectations of increase in the occupancy of each component of the state vector in a unit time, given that the Markov chain is at state $\mathbf{n}$, must all be zero for $\mathbf{n}$ to be an equilibrium point. Applying this principle to the Markov chain represented by Fig 1, then equating to zero the conditional expectations of increase in the number of stations in each mode in a slot time, given that the system is at state **n**, the following set of balance equations is obtained.

$$n_{00}p = n_{0M}(1 - Mp) \tag{2}$$

$$n_{i1}p = n_{i0}p(1 - S') + n_{i+1,0}(1 - p)(1 - S') \quad \text{for } 0 \leq i \leq J - 1 \tag{3}$$

$$n_{J1} = n_{J0}p(1 - S') \tag{4}$$

$$n_{i0}[1 - (1 - p)S'] = n_{iM}(1 - Mp) + n_{i-1,M}Mp + n_{i-1,0}pS' \quad \text{for } 1 \leq i \leq J - 1 \tag{5}$$

$$n_{J0}(1 - S') = n_{JM} + n_{J-1,M} + n_{J-1,0}pS' \tag{6}$$

$$n_{ij} = n_{ij-1} \quad \text{for } 0 \leq i \leq J, 2 \leq j \leq M \tag{7}$$

In the above, the left hand side of the equations represents the expected number of stations leaving the corresponding mode, and the right hand side the expected number of stations entering the corresponding mode, both per slot time, given that the system is at state **n**.

In addition to the balance equations (2) - (7), we have the constraint of a finite number of stations

$$N = \sum_{i=0}^{J} \sum_{j=0}^{M} n_{ij} \tag{8}$$

Note that one of the balance equations (2) - (7) will be linearly dependent on the others under the above constraint.

The initial aim will be to solve the equations (2) - (8) simultaneously to obtain the equilibrium point $\mathbf{n} = \mathbf{n}_e$, under the assumption that the components of the state vector are real rather than integer valued quantities. Then if $S(\mathbf{n})$ represents the system throughput at state **n**, in units of packets per slot time (mean packet transmission time), then since we are effectively replacing the stationary state probability distribution of a Markov chain by a unit impulse located at $\mathbf{n} = \mathbf{n}_e$, the mean value of throughput will be given by

$$E[S(\mathbf{n})] = \int S(\mathbf{n})\delta(\mathbf{n} - \mathbf{n}_e)d\mathbf{n} = S(\mathbf{n}_e) \tag{9}$$

That is, the mean value of throughput can be approximated by its value at the equilibrium point. In order to simplify subsequent notation, we shall write $S = S(\mathbf{n})$ Using this notation, then we have

$$S = \frac{1}{M} \sum_{i=0}^{J} \sum_{j=1}^{M} n_{ij} \tag{10}$$

which, using equations (7) and (8) can be expressed in the simpler form

$$S = \sum_{i=0}^{J} n_{i1} \tag{11}$$

In terms of this quantity, the constraint equation (8) becomes

$$N = \sum_{i=0}^{J} n_{i0} + SM \tag{12}$$

Then using equations (6) and (7) we have

$$n_{j0}(1 - S') = n_{j1} + n_{j-1,1}Mp + n_{j-1,0}pS' \tag{13}$$

Now, using equations (3), (4), (5), (7) and (13) and writing

$$A = \frac{(1 - p)(1 - Mp)(1 - S')}{p[S'(1 - Mp) + Mp]} \tag{14}$$

we find that, in general

$$n_{J-1,0} = n_{J-i+1,0}A \quad \text{for } 1 \leq i \leq J, \tag{15}$$

and

$$n_{J-i,1} = n_{J-i,0}\frac{p}{1 - Mp} \tag{16}$$

Using equation (15) recursively, then

$$n_{J-i,0} = n_{J0}A^i \quad \text{for } 1 \leq i \leq J \tag{17}$$

which is clearly also valid for $i = 0$. Rearranging this we have

$$n_{i0} = n_{J0}A^{J-i} \quad \text{for } 0 \leq i \leq J \tag{18}$$

Next substituting from equation (17) in (16) then

$$n_{J-i,1} = n_{j0}\frac{p}{1 - Mp}A^i \quad \text{for } 1 \leq i \leq J \tag{19}$$

Using equations (4) and (19) in equation (11) and summing the resulting geometric series

$$S = \begin{cases} pn_{J0}\left[(1 - S') + \frac{A}{1-Mp}\left(\frac{1-A^J}{1-A}\right)\right] & A \neq 1 \\ pn_{J0}\left[(1 - S') + \frac{J}{(1-Mp)}\right] & A = 1 \end{cases} \tag{20}$$

Finally, using equation (18) in equation (12), and again summing the geometric series

$$N = \begin{cases} SM + \left(\frac{1-A^{J+1}}{1-A}\right)n_{J0} & A \neq 1 \\ SM + (J + 1)n_{J0} & A = 1 \end{cases} \tag{21}$$

At this point equations (20) and (21) can be solved numerically for S, provided that we can obtain an expression for S', the probability that a station in any of the modes $(i, 0)$, $0 \leq i \leq J$, finds the next passing slot full. To calculate S' we focus attention on a single station in any of the modes in question, and note that slots passing this station can only be occupied by packets from the other $N - 1$ stations. Now let q be the stationary probability that a station occupies a slot with one of its packets, and let X be a random variable representing the number of slots occupied (by $N-1$ stations) out of M. Then since we have assumed stations behave independently, (an assumption that will certainly be true at small values of p when there will be little interaction between stations), X will have a binomial distribution with parameters $(N - 1, q)$, conditioned on the interval $0 \leq X \leq M$. That is

$$P(X = x) = \begin{cases} \frac{1}{R}\binom{N-1}{x}q^x(1 - q)^{N-1-x} & x = 0, 1, \ldots, M \\ 0 & \text{elsewhere} \end{cases} \tag{22}$$

where

$$R = \sum_{i=0}^{M} \binom{N-1}{i} q^i (1-q)^{N-1-i} \tag{23}$$

Now we require the expected number of slots occupied out of M. The mean value of this random variable has been calculated by Harrus [5] as

$$E[X] = (N-1)q - \frac{(1-q)(M+1)}{R} \binom{N-1}{M+1} q^{M+1}(1-q)^{N-1-(M+1)} \tag{24}$$

Now S' is given by

$$S' = \frac{E[X]}{M} \tag{25}$$

Noting that

$$q = \frac{1}{N} \left[\sum_{i=0}^{J} \sum_{j=1}^{M} n_{ij} \right] \tag{26}$$

then from equation (10)

$$q = \frac{MS}{N} \tag{27}$$

An examination of equation (24) shows that for practical values of N and M the second term on the right hand side is small, and little accuracy is lost by ignoring it. Thus S' can be approximated as

$$S' = \frac{(N-1)}{M} q \tag{28}$$

which, using equation (27) gives

$$S' = \frac{(N-1)}{N} S \tag{29}$$

This is an intuitively satisfying result in that it says that the throughput a station observes due to the other $N-1$ stations is simply the total throughput scaled by a factor $(N-1)/N$. In effect, we are approximating a conditional binomial distribution with an unconditioned one having the same parameters $N-1$ and q, on the basis that the right tail of the distribution, where $X > M$, is small and can be discarded without too much loss of accuracy.

Thus, using equation (29) for S' in equations (14) and (20), the latter can now be solved numerically for S, where $0 \leq S \leq 1$. It should be noted that expansion of these equations, in general, results in a polynomial of degree $J + 2$ in S. Since S is a normalised quantity (each slot cannot carry more than one packet at a time) the solution we seek is in the interval $[0, 1]$. We thus proceed by using the equations to obtain a fixed point equation of the form $S = f(S)$, and then solve this for $0 \leq S \leq 1$ by iteration, using some suitable method (such as the bisection method, for example). That there is a unique root of the polynomial in the interval $[0, 1]$ has not been mathematically proved, although the results obtained so far have always supported this uniqueness.

4   Other Performance Measures

In addition to the throughput, the other key performance measures are the mean packet access delay and the probability of buffer overflow. We shall now proceed to evaluate these.

The mean packet access delay, D, can be obtained using Little's result (Little, [8]) by first calculating the mean number of packets stored in the system that are waiting for access to the ring. Denoting this number by Q, then

$$Q = \sum_{i=1}^{J} \left(i \sum_{j=0}^{M} n_{ij} \right) \tag{30}$$

Using equations (4), (7), (18) and (19), and (30), after some manipulation we find

$$Q = \frac{n_{j0}}{(1 - Mp)} \sum_{i=0}^{J-1} iA^{J-1} + Jn_{J0}[1 + Mp(1 - S')] \tag{31}$$

Summing the series, then

$$Q = \begin{cases} \frac{n_{J0}}{(1-Mp)} \left[\frac{(A^{J-1}-1)-(J-1)(1-A^{-1})}{(1-A^{-1})^2} \right] + Jn_{J0}[1 + Mp(1 - S')] & A \neq 1 \\ n_{J0}J \left[\frac{(J-1)}{2(1-Mp)} + 1 + Mp(1 - S') \right] & A = 1 \end{cases} \tag{32}$$

Finally, using Little's result, the mean packet access delay is given by

$$D = \frac{Q}{S} \tag{33}$$

We next find the probability of buffer overflow, B, defined as the probability that a newly generated packet finds a full buffer. Noting that a station in mode $(J, 0)$ will only lose a new packet if the next passing slot is full, we have

$$B = \frac{1}{N} \left[n_{J0}S' + \sum_{j=1}^{M} n_{Jj} \right] \tag{34}$$

Using equations (7) and (29), then

$$B = \frac{n_{J0}}{N} \left[\frac{S(N-1)(1 - Mp)}{N} + Mp \right] \tag{35}$$

5   Results and Discussion

Performance results obtained from the model for throughput, mean packet access delay, and probability of buffer overflow are given respectively in Figs. 2, 3 and 4, where each quantity has been plotted against p, the packet generation probability (arrival probability), for a system with $N = 16$ stations, and $M = 1$ and 16 slots. These latter two quantities are respectively the minimum and maximum numbers of slots that can be used in the model under the assumption that $M \leq N$. Performance results for intermediate values of M fall between the two extremes considered. The results are compared with those obtained from an equivalent simulation study, from

159

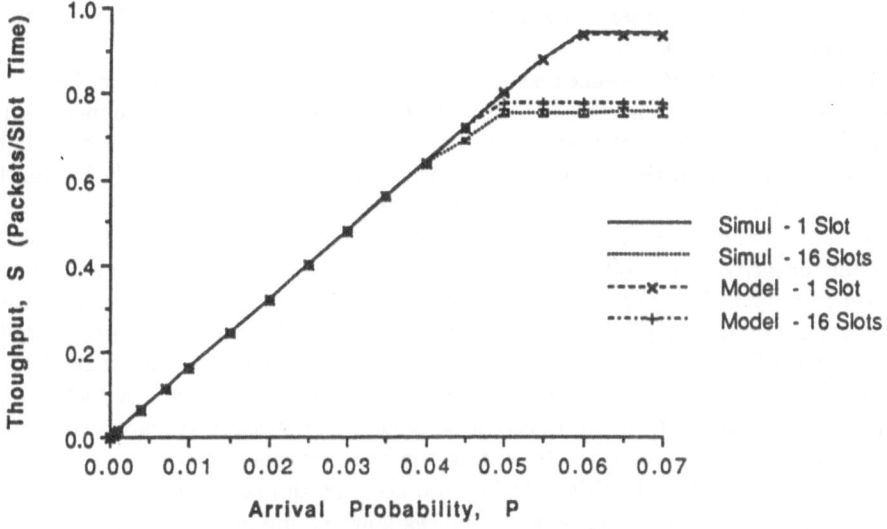

fig. 2   Throughput against P for N=16, J=100

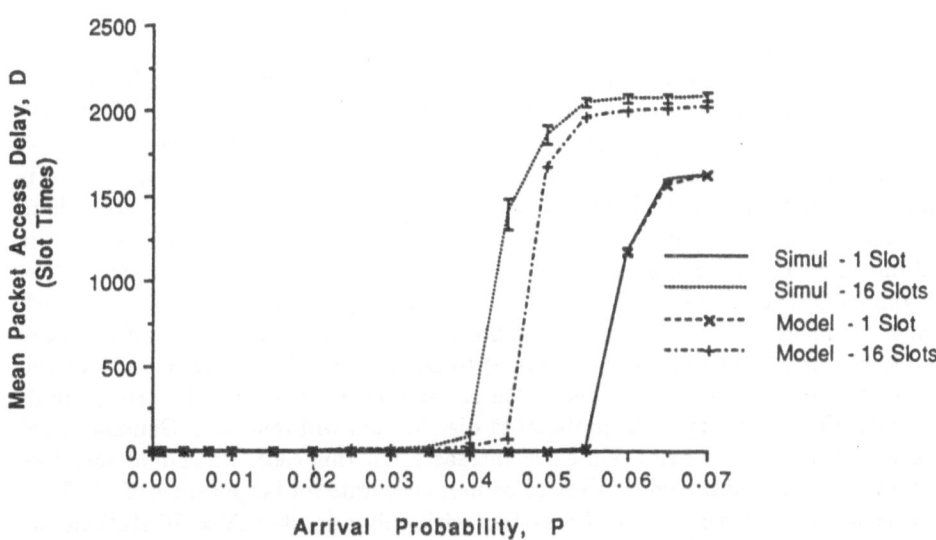

fig. 3   Mean Packet Access Delay against P for N=16, J=100

which 95% confidence intervals were obtained and are also shown. A buffer size of $J = 100$ is used throughout.

In all cases the performance results obtained from the model with $M = 1$ slot are extremely close to the simulation results. With $M = 16$ slots the performance results obtained are again close at loads up to saturation, after which the model tends to deteriorate and overestimate the ring's performance. This could partly be due to the assumption that was made to simplify the model, that a station cannot generate more than one new packet during its own transmission period. It is likely however that the majority of the error is due to the existence of quasi-stable states, which are known to reduce the system's transmission capacity. Although a discussion of this topic is outside the scope of the present article, it was shown by Falconer, Adams, and Whalley [3] that these states are very prevalent on multiple slot rings at high network loadings, when the probability of the station buffers being empty is very small. To take such states into account it would be necessary to explicitly model the ordering of slots in relation to stations on the ring, and it seems very doubtful that any model could take this ordering into account and still remain tractable.

The final performance curve, Fig 5, shows how the model can be used to assess the effects of buffer size on the probability of buffer overflow. Clearly, from Fig 5, once the buffer size exceeds a modest value, in this case about $J = 10$, then this has very little effect in reducing the probability of buffer overflow for a given value of p. This feature of the model, to predict packet loss characteristics, is one that is unavailable in any previous modelling study of the slotted ring.

With regard to the modelling technique used, it should be noted that equation (2) was not used in solving the model, since this is linearly dependent on the other balance equations. In this context, the state vector of equation (1) might be redefined by omitting one of the components (n_{00} say) since this can always be obtained in terms of the other components by using the constraint equation (8).

Concerning the general question of the accuracy of EPA, since the method is based on replacing the stationary state probability distribution of a Markov chain by a unit impulse located at an equilibrium point, then we should expect the results obtained by EPA to be more accurate if the stationary state probability distribution is symmetrical about the equilibrium point. For the type of Markov chain considered, which models the interactions of N identical entities, the components of the state vector can be considered (approximately) to be the sum of N independent random variables, each of which takes the value either 0 or 1. Then by the central limit theorem, the stationary state probability distribution will tend to a Gaussian (and hence symmetrical) distribution as N increases. In this case, we should therefore expect the results obtained by EPA to be more accurate for large values of N. Since in this article we have examined a system with only a modest $N = 16$ stations, we should correspondingly expect the results of the model to become more accurate as N is increased.

Finally, the fact that EPA is dependent on the network stations being statistically identical is obviously somewhat restrictive. It has recently been shown however that EPA is equivalent to the so called fixed point approximation method under certain constraints (Woodward [12]). Since this latter method considers single network stations in isolation, then it seems likely that EPA can be extended to the case of non-identical stations via this equivalence.

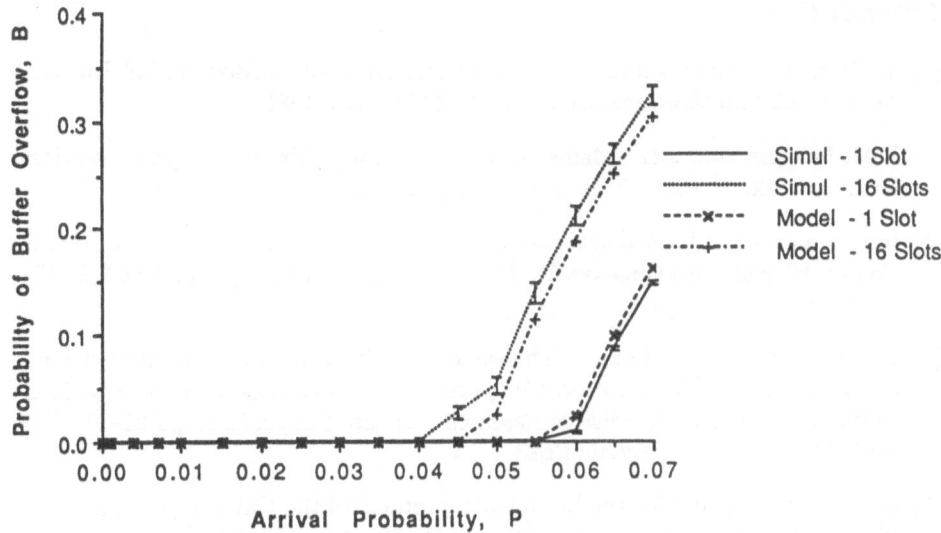

fig. 4   Probability of Buffer Overflow against P for N=16, J=100

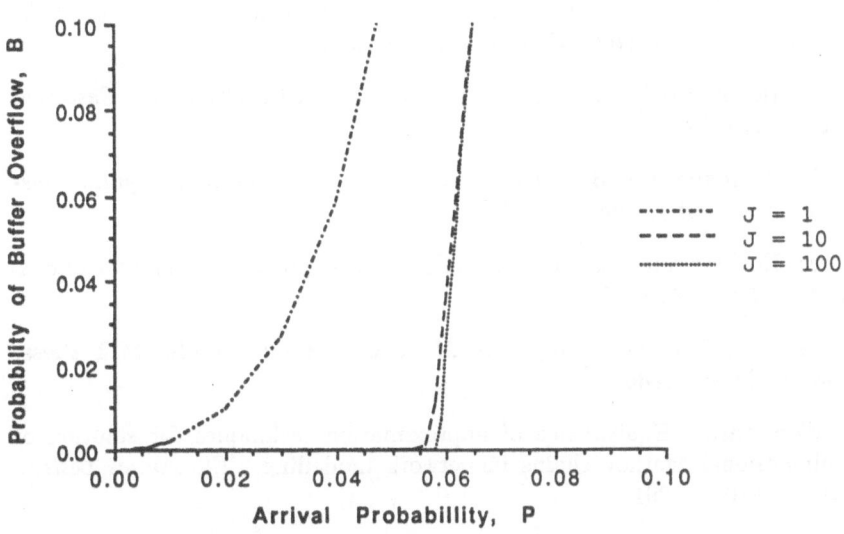

Fig. 5   Probability of Buffer Overflow against P for N=16 and various J

References

[1] W. Bux. Local area subnetworks: A performance comparison. *IEEE Transactions on Communications*, COM-29(10):1465–1473, 1981.

[2] R.M. Falconer and J.L. Adams. Orwell: A protocol for an integrated services local network. *British Telecom Technology Journal*, 3(4):27–35, 1985.

[3] R.M. Falconer, J.L. Adams, and G.M. Whalley. A simulation study of the Cambridge ring with voice traffic. *British Telecom Technology Journal*, 3(2):85–91, 1985.

[4] D.J. Farber and K.C. Larson. The system architecture of the distributed computer system. In J. Fox, editor, *Computer Communications Networks and Teletraffic*, volume 22 of *Microwave Research Institute Symposia*, pages 21–27, New York, NY, 1972. Polytechnic Press.

[5] G. Harrus. A model for the basic block protocol of the Cambridge ring. *IEEE Transactions on Software Engineering*, SE-11(1):130–136, 1985.

[6] A.R. Kamal and V.C. Hamacher. Approximate analysis of non-exhaustive multi-server polling systems with applications to local area networks. *Computer Networks and ISDN Systems*, 17(1):15–27, 1989.

[7] P.J.B. King and I. Mitrani. Modelling a slotted ring local area network. *IEEE Transactions on Computers*, C-36(5):554–561, 1987.

[8] J.D.C. Little. A proof of the queuing formula $L = \lambda W$. *Operations Research*, 9(3):383–387, 1961.

[9] R.M. Needham and A.J. Herbert. *The Cambridge Distributed Computing System*. Addison-Wesley, London, 1982.

[10] J. Pierce. How far can data loops go? *IEEE Transactions on Communications*, COM-20(3):527–530, 1972.

[11] S. Tasaka. *Performance Analysis of Multiple Access Protocols*. MIT Press, Cambridge, Mass., 1986.

[12] M.E. Woodward. Equivalence of approximation techniques for solution of multidimensional Markov chains in network modelling. *Electronics Letters*, 27(12):1019–1021, 1991.

MEM for Arbitrary Exponential Open Networks with Blocking and Multiple Job Classes [*]

Demetres D Kouvatsos
Spiros G Denazis

Computer Systems Modelling Research Group,
University of Bradford, Bradford,
United Kingdom

Panagiotis H Georgatsos

Telecommunications Division
ALFA SAI, 72-74 Salaminos Street
Kallithea, Athens, Greece

Abstract

The Maximum Entropy Method (MEM) is used to approximate the joint stationary queue length distribution of an $M/M/1/N$ queueing system with finite capacity, N, $R(R > 1)$ classes of jobs under an unrestricted buffer sharing scheme and mixed service disciplines drawn from FCFS, LCFS-NPR, LCFS-PR and PS rules. The marginal and aggregate ME queue length distributions and the associate blocking probabilities per class are also determined. These ME results in conjunction with the first moment of the effective flow are used, as building blocks, in order to establish a new product-form approximation for arbitrary exponential multi-class open queueing networks under repetitive-service (RS) blocking with random destination (RD). Numerical experiments illustrate the credibility of the ME approximations in relation to simulation.

1 Introduction

Queueing network models (QNMs) with finite capacity and multi-classes of jobs are widely recognised as powerful tools for analysing computer and communication systems and optimising their performance. Exact closed-form solutions for this type of QNMs are not generally attainable except for some special cases such as two-station cyclic queues and reversible networks. As a consequence, numerical techniques and analytic approximations are used for the study of general QNMs with finite capacity.

---

[*]This work is supported in part by the Science and Engineering Research Council (SERC), UK, under grant GR/F29271

A useful bibliography on the subject has been compiled by Perros [1]. Moreover, two comprehensive surveys on open and closed QNMs under various blocking mechanisms have been carried out, respectively, by Perros [2] and Onvural [3]. However, not much work has been done so far on QNMs with multi-classes of jobs and general service times. Akyildiz and von Brand [4–6] analysed open, closed and mixed QNMs with multi-classes and reversible routing under repetitive-service (RS) blocking. Onvural [7] has shown that some product form stationary distributions for special single class networks hold also with multi-classes and BCMP type stations. More recently, Onvural and Perros [8] developed a fairly accurate approximation procedure for open multi-class tandem queueing networks with finite capacity and Coxian-2 interarrival and service times, under non-preemptive head-of-line priority rule.

In this paper earlier work by Kouvatsos and Xenios [9] and Kouvatsos and Denazis [10] on arbitrary single-class QNMs under RS blocking is extended to the multi-class case. A new product-form approximation is proposed for arbitrary exponential multi-class open QNMs with single-server stations of finite capacity under RS blocking with random destination (RD) and mixed service disciplines drawn from FCFS (first-come-first-served), LCFS-NPR (last-come-first-served non-preemptive), LCFS-PR (last-come-first-served preemptive) and PS (Processor Sharing) rules. Entropy maximisation implies a decomposition of the open QNM into individual multi-class $M/M/1/N$ queueing systems with finite capacity, N, under revised service-time distributions and censored interarrival processes. The multi-class $M/M/1/N$ queueing system with distinct exponential interarrival and service times per class is used, as a building block, in the solution process.

The ME analysis of an $M/M/1/N$ queue is presented in Sect. 2. The ME product-form approximation for arbitrary open QNMs and the associated MEM algorithm are given in Sect. 3. Numerical validation experiments are carried out in Sect. 4. Concluding remarks follow in Sect. 5.

2 The multi-class $M/M/1/N$ queue with unrestricted sharing

Consider a stable $M/M/1/N$ queue with finite capacity, N, a single server, $R(R > 1)$ job classes, exponential interarrival or service times and mixed scheduling disciplines drawn from FCFS, LCFS-NPR, LCFS-RP and PS rules. It is assumed that buffer management employs an unrestricted sharing scheme, where buffers are allocated to jobs according to a given scheduling rule while no discrimination is made on the basis of job class. Moreover, the arrival process per class is considered to be "censored" i.e., arriving jobs of class $i, i = 1, 2, \ldots, R$ are turned away when all N buffers are full.

2.1 Notation

With reference to the $G/G/l/N$ queue, let at any given time $\underline{k} = (k_1, k_2, \ldots, k_R)$ be a system state, where k_i $(k_i \geq 0)$ is the number of jobs of class i, $i = 1, 2, \ldots, R$, S be a system state where jobs are positioned under an ordered arrangement, Q be the set of all feasible states S. For each state S and class i, $i = 1, 2, \ldots, R$ let $n_i(S)$, be the number of class i jobs present in state S and $s_i(S)$, $f(S)$ be auxiliary functions given by

$$s_i(S) = \begin{cases} 1 & \text{if the job in service is of class } i, \\ 0 & \text{otherwise} \end{cases}$$

$$f(S) = \begin{cases} 1 & \text{if } \sum_{j=1}^{R} n_j(S) = N, \\ 0 & \text{otherwise} \end{cases}$$

Moreover, let Λ_i be the rate of the overall arrival process, μ_i be the rate of the service process and π_i be the blocking probability that an arrival of class i finds the queue full (NB. under exponential assumptions this is identical to the aggregate full state probability $p(N)$). Finally, let $p(S)$ and $p(k_1, k_2, \ldots, k_R)$ be the stationary state probabilities.

2.2 The joint ME solution

Motivated by earlier applications of the ME principle for the approximate analysis of single-class finite capacity queues and networks (c.f., Kouvatsos [11], Kouvatsos and Xenios [9]) and multi-class infinite capacity queues (c.f., Kouvatsos and Tabet-Aouel [12]), it is assumed that the following mean value constraints about the state probabilities $\{p(S)\}$, $S \in Q$, are known to exist:

(i) Normalisation,

$$\sum_{S \in Q} p(S) = 1. \tag{1}$$

(ii) Utilisation, ν_i, $0 < \nu_i < 1$, $i = 1, 2, \ldots, R$,

$$\sum_{S \in Q} s_i(S) p(S) = \nu_i. \tag{2}$$

(iii) Mean queue length, $\langle n_i \rangle$, $\nu_i < \langle n_i \rangle < N$, $i = 1, 2, \ldots, R$,

$$\sum_{S \in Q} n_i(S) p(S) = \langle n_i \rangle. \tag{3}$$

(iv) Full buffer state probability, φ, $0 < \varphi < 1$, $i = 1, 2, \ldots, R$,

$$\sum_{S \in Q} f(S) p(S) = \varphi \tag{4}$$

satisfying the flow-balance condition

$$\Lambda_i(1 - \pi_i) = \mu_i \nu_i. \tag{5}$$

The form of the state probability distribution, $p(S)$, can be characterised by maximising the entropy functional

$$H(p) = -\sum_{S \in Q} p(S) \log p(S), \tag{6}$$

subject to constraints (1) - (4). By employing the method of Lagrange's undetermined multipliers the following solution is obtained

$$p(S) = \frac{1}{Z} \prod_{i=1}^{R} g_i^{s_i(S)} x_i^{n_i(S)} y^{f(S)}, \qquad \forall S \in Q. \tag{7}$$

Defining the sets

$$S_0 = \{S | S \in Q : s_i(S) = 0, \ \forall i = 1, 2, \dots, R\}$$

$$Q_i = \{S | S \in Q : s_i(S) = 1\}, \qquad i = 1, 2, \dots, R$$

$$Q_{i;k_1,\dots,k_R} = \{S \in Q_i : n_j(S) = k_j \ \& \ k_i \geq 1, \ j = 1, 2, \dots, R\}, \qquad i = 1, 2, \dots, R,$$

it is implied, after some manipulation, that the ME joint state probability distribution is given by

$$p(S_0) = p(0,\dots,0) = \frac{1}{Z}, \tag{8}$$

$$
\begin{aligned}
p(k_1,\dots,k_R) &= \sum_{i=1}^{R} Prob\left(Q_{i;k_1,\dots,k_R}\right) \\
&= Z^{-1} \frac{\left(\sum_{j=1}^{R} k_j - 1\right)!}{\prod_{j=1}^{R} k_j!} \left(\prod_{j=1}^{R} x_j^{k_j}\right) \left(\sum_{i=1}^{R} k_i g_i\right) y^{\delta\left(\sum_{j=1}^{R} k_j\right)}, \tag{9}
\end{aligned}
$$

where

$$\delta\left(\sum_{j=1}^{R} k_j\right) = \begin{cases} 1, & \text{if } \sum_{j=1}^{R} k_j = N, \\ 0, & \text{otherwise.} \end{cases} \tag{10}$$

2.3   The marginal ME solution

The marginal state probabilities per class $p_i(k)$, $i = 1, 2, \dots, R$, $k = 0, 1, \dots, N$, can be determined by using their standard definition in terms of the joint state probabilities (8) - (9). To this end, after some manipulation, the following formulae are obtained:

$$p_i(0) = \frac{1}{Z} \left\{ 1 + \left(\sum_{i \neq j=1}^{R} x_j g_j\right) \left(\frac{\left(1 - \bar{x}_i^{N-1}\right)}{1 - \bar{x}_i} + y \bar{x}_i^{N-1}\right) \right\} \tag{11}$$

$$p_i(k) = \frac{1}{Z}x_i^k \left\{ g_i \sum_{n=0}^{N-k-1} \binom{n+k-1}{k-1} \bar{x}_i^n \right.$$

$$+ \left(\sum_{i \neq j=1}^{R} x_j g_j \right) \sum_{n=0}^{N-k-2} \binom{n+k}{k} \bar{x}_i^n \tag{12}$$

$$+ y g_i \binom{N-1}{k-1} \bar{x}_i^{N-k}$$

$$\left. + y \left(\sum_{i \neq j=1}^{R} x_j g_j \right) \binom{N-1}{k} \bar{x}_i^{N-k-1} \right\}, \qquad k = 1, 2, \ldots, N-1,$$

where $\bar{x}_i = X - x_i$, $X = \sum_{i=1}^{R} x_i$ and

$$p_i(N) = \frac{1}{Z} g_i x_i^N y. \tag{13}$$

Note that the analytic expressions (11) - (13) of the marginal probabilities are rather complex and computationally expensive. However, by using Pascal's triangle inequality and carrying out laborious but not complex operations, the following recursive expressions are obtained:

$$Z^{(1)} = 1 + \frac{\rho(1-X)}{1-\rho}y, \tag{14}$$

$$Z^{(N)} = \frac{1-X}{1-\rho} + XZ^{N-1}, \qquad N \geq 2, \tag{15}$$

$$p^{(N)}(S_0) = 1/Z^{(N)}. \tag{16}$$

For $i = 1, \ldots, R$ and $N \geq 2$,

$$p_i^{(1)}(0) = 1 - \frac{1}{Z^{(1)}} \left(\frac{1-X}{1-\rho} \right) \rho_i y, \tag{17}$$

$$p_i^{(N)}(0) = \frac{Z^{(N-1)}}{Z^{(N)}} \bar{x}_i p_i^{(N-1)}(0) + \frac{1}{Z^{(N)}} A_i, \tag{18}$$

$$p_i^{(1)}(1) = \frac{1}{Z_0^{(1)}} x_i g_i y, \tag{19}$$

$$p_i^{(N)}(1) = \frac{Z^{(N-1)}}{Z^{(N)}} \left(\bar{x}_i p_i^{(N-1)}(1) + x_i p_i^{(N-1)}(0) \right) - \frac{1}{Z^{(N)}} x_i (1 - g_i), \tag{20}$$

$$p_i^{(N)}(k) = \frac{Z^{(n-1)}}{Z^{(n)}} \left(\bar{x}_i p_i^{(N-1)}(k) + x_i p_i^{(N-1)}(k-1) \right), \qquad k = 2, \ldots, N, \tag{21}$$

(N.B. $p_i^{(N-1)}(N) = 0$) where $\bar{x}_i = X - x_i$, and $A_i = (1 - \bar{x}_i) + \frac{1-X}{1-\rho}(\rho - \rho_i)$.

2.4 The Langrangian coefficients $\{x_i\}, \{g_i\}, \{y\}$

By making asymptotic connections to an infinite capacity queue as $N \to +\infty$ and assuming (a) $\rho_i < 1$ and $X < 1$ and (b) $\{x_i\}, \{g_i\}$ and $\{y\}$ are invariant to the buffer capacity size N, it can be established under exponential interarrival and service time distributions per class that

$$x_i = \frac{\bar{n}_i - \rho_i}{\bar{n}} \tag{22}$$

$$g_i = \frac{(1 - X)\rho_i}{(1 - \rho)x_i} \tag{23}$$

Moreover, by using flow balance condition (5), it follows that

$$y = \frac{1 - \rho}{1 - X}, \tag{24}$$

where $\bar{n} = \sum_{i=1}^{R} \bar{n}_i$ and $\bar{n}_i$ is the asymptotic mean queue length determined by (c.f., Kouvatsos and Tabet-Aouel [12])

$$\bar{n}_i = \begin{cases} \rho_i + \frac{\Lambda_i}{1-\rho} \sum_{j=1}^{R} \frac{\rho_j^2}{\Lambda_j}, & \text{if FCFS or LCFS-NP scheduling rule} \\[2mm] \frac{\rho_i}{1-\rho} & \text{if LCFS-PR scheduling rule} \\[2mm] \rho_i \left\{ 1 + \frac{1}{1-\rho} \sum_{j=1}^{R} \frac{h_j}{h_i} \rho_j \right\}, & \text{if PS scheduling rule} \end{cases}$$

where $\rho_i = \Lambda_i/\mu_i, \rho = \sum_{i=1}^{R} \rho_i, \Lambda = \sum_{i=1}^{R} \Lambda_i$, and $\{h_i, i = 1, \ldots, R\}$ is a set of known discriminatory weights that impose differential treatment to different classes [15].

2.5 The aggregate ME solution

The aggregate state probability, $p(n), n = 0, 1, \ldots, N$, can be completely specified by making use of (8), (9) and (22) - (24), namely,

$$p(0) = p(S_0) = Z^{-1} \tag{25}$$

$$p(n) = \sum_{\underline{k} \in A_n} p(\underline{k}) = Z^{-1} \frac{\rho(1 - X)}{1 - \rho} X^{n-1}, \qquad 1 \leq n \leq N - 1, \tag{26}$$

where

$$A_n = \left\{ \underline{k} : 0 \leq k_j \leq N, \ j = 1, \ldots, R \text{ and } \sum_{j=1}^{R} k_j = n \right\},$$

$$\rho = \sum_{j=1}^{R} \rho_j, \qquad \rho_j = \Lambda_j/\mu_j, \text{ and}$$

$$p(N) = Z_N^{-1} \rho \frac{1 - X}{1 - \rho} X^{N-1} y. \tag{27}$$

Note that details of proofs for expressions (8) - (9) can be found in Kouvatsos and Georgatsos [13].

3 Arbitrary open multi-class networks with unrestricted sharing and RS blocking with RD

Consider an arbitrary open QNM at equilibrium containing M single-server queueing stations with finite capacity N_ℓ, $\ell = 1, \ldots, M$. The network operates under an unrestricted buffer sharing scheme, RS blocking with RD and FCFS, LCFS-NPR, LCFS-PR and PS scheduling rules.

3.1 Notation

For each queueing station ℓ or m, $\ell, m = 1, 2, \ldots, M$, the notation of Sect. 2 applies and in addition, let

$\Lambda_{O;\ell,i}$ be the overall arrival rate of class i to queue ℓ, $i = 1, 2, \ldots, R$, $\ell = 1, 2, \ldots, M$.

$a_{\ell,i;m,j}$ be the transition probability that a class i completer from queue ℓ attempts to join queue m as class $j, \ell = 1, 2, \ldots, M, m = 0, \ldots, M, i, j = 1, 2, \ldots, R$.

$\pi_\ell = p_\ell(N_\ell)$ be the blocking probability that an external arriver or a completer in the network is blocked by queue ℓ, $\ell = 1, 2, \ldots, M$.

$\pi_{c\ell,i}$ be the blocking probability that a class i completer from queue ℓ (in the first or subsequent attempts) is blocked.

3.2 The joint ME solution

The state of the network at any given time can be described by an integer valued vector $\underline{k} = (\underline{k}_1, \underline{k}_2, \ldots, \underline{k}_M)$ where $\underline{k}_\ell = (k_{\ell,1}, \ldots, k_{\ell,R})$ and $k_{\ell,i}, i = 1, 2, \ldots, R$ is the number of class i jobs at queue ℓ, $\ell = 1, 2, \ldots, M$. Let $p(\underline{k})$ be the joint stationary probability that the queueing network is at state k.

The form of the ME solution $p(\underline{k})$, subject to normalisation and the marginal constraints of the type (2) - (4), can be clearly established in terms of the product-form approximation

$$p(\underline{k}) = \prod_{\ell=1}^{M} p_\ell(\underline{k}_\ell), \qquad (28)$$

where $p_\ell(\underline{k}_\ell)$ is the marginal ME solution expressed by (9). The ME solution implies an $M/M/1/N_\ell$ queue-by-queue decomposition with revised interarrival and service processes. Assuming a Poisson arrival process at each queue, the modified MEM algorithm of Kouvatsos and Denazis [10] (applied to the single class case) can be extended to produce a MEM algorithm for open multi-class networks.

3.3 The MEM algorithm

The MEM algorithm for arbitrary multi-class open QNMs with single servers exponential interarrival and service times, class switching, unrestricted buffer sharing, RS blocking under RD and FCFS, LCFS-NPR, LCFS-PR and PS scheduling rules can be described as follows:

Begin

    Inputs
    $\{M, R, N_\ell, \mu_{\ell,i}, \Lambda_{O;\ell,i};\ \ell = 1, 2, \ldots, M;\ i = 1, 2, \ldots, R,$ FCFS,
    LCFS-NPR, LCFS-PR, PS scheduling rules, $\{a_{\ell,i;m,j}\}$,
    $\ell = 1, 2, \ldots, M;\ m = 0, 1, \ldots, M;\ i, j = 1, 2, \ldots, R\}$.

Step 1   **Feedback Correction**
    For each queue $\ell, \ell = 1, 2, \ldots, M,$
    For each class $i, i = 1, 2, \ldots, R,$ (with $a_{\ell,i;\ell,i} > 0$)
    $\mu_{\ell,i} \leftarrow \mu_{\ell,i}(1 - a_{\ell,i;\ell,i})$
    $a_{\ell,i;m,j} \leftarrow a_{\ell,i;m,j}/(1 - a_{\ell,i;\ell,i})$,
    (for all $m \neq \ell, m = 0, 1, \ldots, M$ and $j \neq i$, $j = 1, \ldots, R$)
    $a_{\ell,i;\ell,i} \leftarrow 0;$

Step 2   **Initialisation**
    For each queue $\ell = 1, 2, \ldots, M,$ set
    $\pi_\ell = P_\ell(N_\ell) \leftarrow 0.5;$

Step 3   **Solve System of Non-Linear Equations $\{M_\ell\}$**
    Using Newton-Raphson (c.f.,(27))

$$\begin{aligned}
\pi_\ell &= p_\ell(N_\ell) \leftarrow \frac{1}{Z}\rho_\ell\left(\frac{1-X_\ell}{1-\rho_\ell}\right)X_\ell^{N_\ell-1}y \\
&= \frac{(1-\rho_\ell)}{1-\rho_\ell^2 X_\ell^{N_\ell-1}}\rho_\ell X_\ell^{N_\ell-1}, \qquad \ell = 1, 2, \ldots, M;
\end{aligned}$$

Step 3.1   **Calculate effective flow transition probabilities $\{\bar{a}_{\ell,i;m,j}\}$**
    $\bar{a}_{\ell,i;m,j} \leftarrow a_{\ell,i;m,j} * (1 - \pi_m)/(1 - \pi_{c\ell,i});$

Step 3.2   **Calculate effective job flow balance equations**
    $\lambda_{\ell,i} \leftarrow \lambda_{O;\ell,i} + \sum_{\ell \neq m=1}^{M}\sum_{j=1}^{R}\bar{a}_{m,j;\ell,i}\lambda_{m,j}$
    $\lambda_{O;\ell,i} \leftarrow \Lambda_{O;\ell,i}(1 - \pi_\ell);$

Step 3.3   **Calculate effective service-rate**
    $\hat{\mu}_{\ell,i} \leftarrow \mu_{\ell,i}(1 - \pi_{c;\ell,i});$

Step 3.4   **Calculate blocking probability $\pi_{c;\ell,i}$**
    $\pi_{c;\ell,i} \leftarrow \sum_{\ell \neq m=1}^{M}\left(\sum_{j=1}^{R}a_{\ell,i;m,j}\right)\pi_m$

Step 3.5   **Calculate overall arrival rate $\Lambda_{\ell,i}$**
    $\Lambda_{\ell,i} \leftarrow \lambda_{\ell,i}/(1 - \pi_\ell);$

    (NB. $\pi_\ell = p_\ell(N_\ell)$ is the marginal stationary probability of
    having a full multi-class $M(\Lambda_{\ell,i})/M(\hat{\mu}_{\ell,i}/1/N_\ell$ queue under
    a specific scheduling discipline drawn from FCFS, LCFS-PR,
    LCFS-NPR and PS rules.)

Step 4 Output

Solve each queueing station $\ell, \ell = 1, 2, \ldots, M$ in isolation as a censored $M/M/1/N_\ell$ queue with multi-classes of jobs with parameters $(\Lambda_{\ell,i}, \hat{\mu}_{\ell,i})$ under a given scheduling discipline.

End

4 Numerical results

An indication of the credibility of the proposed MEM is demonstrated via three typical simulation examples. In particular Example 1 refers to an $M/M/1/N$ queue with three classes of jobs under FCFS scheduling, involving the aggregate mean queue length ($\langle n \rangle$), response time (R), idle and full probabilities $\{p(0), p(N)\}$ (Table 1.1) and the marginal utilizations $\{\nu_j\}$, mean queue lengths $\{\langle n_j \rangle\}$, response times $\{R_i\}$ and idle probabilities $\{p_i(0)\}$ (Table 1.2). Example 2 is an open central server model with three stations visited by two classes of jobs where each station operates under a FCFS scheduling discipline. Finally, Example 3 is a tandem network with feedback and visited by three classes of jobs and each station operates under a LCFS-PR scheduling discipline. For each station ℓ the following statistics are presented: the aggregate mean queue length, $\langle n_\ell \rangle$, response time, R_ℓ, idle and full probabilities $\{p_\ell(0), p_\ell(N_\ell)\}$ (Tables 2.1 and 3.1), marginal utilization, $\nu_{\ell,i}$ mean queue length, $\langle n_{\ell,i} \rangle$ and idle probabilities $\{p_{\ell,i}(0)\}$ (Tables 2.2 and 3.2).

These examples facilitate relative comparisons between MEM solutions and simulation (SIM) results (obtained via QNAP-2 [14]). For validation purposes two kinds of error indicators are used:

1. the percentage difference (%D) which is used throughout Examples 1-3 (aggregate statistics), and

2. the tolerance error (TOL) which is used in Examples 2 and 3 (marginal statistics) and is defined for the marginal mean queue lengths, as the ratio

$$\left| \frac{\text{SIM}(\langle n_{\ell,i} \rangle) - \text{MEM}(\langle n_{\ell,i} \rangle)}{\sum_{\ell=1}^{M} \text{SIM}(\langle n_{\ell,i} \rangle)} \right|$$

whereas for the marginal utilizations and idle probabilities $\{(p_{\ell,i}(0)\}$,

$$\left| \text{SIM}(\nu_{\ell,i}) - \text{MEM}(\nu_{\ell,i}) \right|$$

and

$$\left| \text{SIM}(p_{\ell,i}(0)) - \text{MEM}(p_{\ell,i}(0)) \right|$$

respectively. It can be observed that the ME solutions are very comparable in accuracy with those obtained via simulation.

Example 1: $M/M/1/N$ queue

Raw data: $R = 3$, FCFS, $N = 5$, $\bar{\Lambda} = (2, 1, 3)$, $\bar{\mu} = (3, 4, 5)$

Table 1.1: Aggregate statistics

Measures	SIM	MEM	% D
$\langle n \rangle$	3.5950	3.5750	-0.556%
R	0.9617	0.9533	-0.873%
$p(O)$	0.0490	0.0520	6.122%
$p(S)$	0.3769	0.3750	-0.504%

Table 1.2: Marginal statistics

Class	Measures	SIM	MEM	% D
1	ν_1	0.4170	0.4167	-0.072%
	$\langle n_1 \rangle$	1.299	1.2920	-0.539%
	R_1	1.044	1.0340	-0.956%
	$p_1(0)$	0.2589	0.2529	-2.317%
2	ν_2	0.1575	0.1563	-0.7612%
	$\langle n_2 \rangle$	0.6006	0.5941	-1.082%
	R_2	0.9585	0.9506	-0.824%
	$p_2(0)$	0.5405	0.5438	0.611%
3	ν_3	0.3769	0.3750	-0.504%
	$\langle n_3 \rangle$	1.6960	1.6890	-0.413%
	R_3	0.9080	0.9008	-0.793%
	$p_3(0)$	0.1732	0.1843	6.409%

Example 2: Central server model

Raw data: $M = 3$, $R = 2$, FCFS

Queue 1:
$$\bar{\Lambda}_{0,1} = (1,6,2), \quad \bar{\mu}_1 = (3,4), \quad N_1 = 15,$$
$$a_{11;21} = 1/6, \quad a_{11;22} = 1/6, \quad a_{11;31} = 2/9,$$
$$a_{11;32} = 1/9, \quad a_{12;21} = 1/8, \quad a_{12;22} = 1/8,$$
$$a_{12;32} = 1/8, \quad a_{12;31} = 1/8, \quad a_{12;32} = 1/8,$$
$$a_{11;0} = 1/3, \quad a_{12;0} = 1/2,$$

Queue 2:
$$\bar{\mu}_2 = (5,2) \quad N_2 = 10$$
$$a_{21;11} = 1/2 \quad a_{21;12} = 1/2$$
$$a_{22;11} = 2/3 \quad a_{22;12} = 1/3$$

Queue 3:
$$\bar{\mu}_3 = (2,2) \quad N_3 = 12$$
$$a_{31;11} = 1/5 \quad a_{31;12} = 4/5$$
$$a_{32;11} = 1/5 \quad a_{32;12} = 4/5$$

Table 2.1: Aggregate statistics

Queue	Measures	SIM	MEM	%D
1	$\langle n_1 \rangle$	14.39	14.385	-0.035%
	R_1	4.759	4.766	0.147%
	$p_1(0)$	0	0	0%
	$p_1(15))$	0.625	0.625	0%
2	$\langle n_2 \rangle$	4.332	4.377	1.039%
	R_2	4.658	4.737	1.696%
	$p_2(0)$	0.133	0.138	3.76%
	$p_2(10)$	0.071	0.075	5.634%
3	$\langle n_3 \rangle$	9.339	9.306	-0.353%
	R_3	12.569	12.558	-0.088%
	$p_3(0)$	0.008	0.008	0%
	$p_3(12)$	0.259	0.255	-1.506%

Table 2.2: Marginal statistics

Queue	Class	Measures	SIM	MEM	TOL
1	1	$\nu_{1,1}$	0.6503	0.650	0.0003
		$\langle n_{1,1} \rangle$	8.311($\pm$0.007)	8.341	0.0020
		$p_{1,1}(0)$	0	0	0
	2	$\nu_{1,2}$	0.3497	0.35	0.0003
		$\langle n_{1,2} \rangle$	6.079($\pm$0.006)	6.044	0.0030
		$p_{1,2}(0)$	0.0005	0.0005	0
2	1	$\nu_{2,1}$	0.2905	0.246	0.0445
		$\langle n_{2,1} \rangle$	2.024($\pm$0.02)	2.004	0.0010
		$p_{2,1}(0)$	0.267	0.286	0.0190
	2	$\nu_{2,2}$	0.5765	0.616	0.0400
		$\langle n_{2,2} \rangle$	2.309($\pm$0.025)	2.373	0.0050
		$p_{2,2}(0)$	0.2137	0.197	0.0168
3	1	$\nu_{3,1}$	0.6028	0.603	0.0002
		$\langle n_{3,1} \rangle$	5.669($\pm$0.025)	5.661	0.0005
		$p_{3,1}(0)$	0.0166	0.017	0.0005
	2	$\nu_{3,2}$	0.3894	0.388	0.0014
		$\langle n_{3,2} \rangle$	3.670($\pm$0.02)	3.645	0.0020
		$p_{3,2}(0)$	0.0394	0.040	0.0006

Example 3: Tandem Queues

Raw Data: $M = 4$, $R = 3$, LCFS-PR

Queue 1: $\quad \bar{\Lambda}_{0,1} = (1, 2, 1.5)$, $\quad \bar{\mu}_1 = (5, 5, 2)$, $\quad N_1 = 10$,

$\quad\quad a_{11;21} = 0.3$, $\quad a_{11;22} = 0.3$, $\quad a_{11;23} = 0.4$,

$\quad\quad a_{12;21} = 0.2$, $\quad a_{12;22} = 0.5$, $\quad a_{12;23} = 0.3$,

$\quad\quad a_{13;21} = 0.45$, $\quad a_{13;22} = 0.45$, $\quad a_{13;23} = 0.1$,

Queue 2: $\quad \bar{\mu}_2 = (4, 5, 6)$ $\quad N_2 = 5$

$\quad\quad a_{21;31} = 0.6$ $\quad a_{21;32} = 0.4$

$\quad\quad a_{22;31} = 1$

$\quad\quad a_{23;31} = 0.4$ $\quad a_{23;32} = 0.4$ $\quad a_{23;33} = 0.2$

Queue 3: $\quad \bar{\mu}_3 = (3, 2, 4)$ $\quad N_3 = 10$

$\quad\quad a_{31;41} = 0.4$ $\quad a_{31;42} = 0.6$

$\quad\quad a_{32;41} = 0.3$ $\quad a_{32;42} = 0.2$ $\quad a_{32;43} = 0.5$

$\quad\quad a_{33;41} = 1$

Queue 4: $\quad \bar{\mu}_4 = (3, 2, 3)$ $\quad N_4 = 5$

$\quad\quad a_{41;11} = 0.05$ $\quad a_{41;12} = 0.1$ $\quad a_{41;13} = 0.05$

$\quad\quad a_{41;0} = 0.08$ $\quad a_{42;11} = 0.15$ $\quad a_{42;12} = 0.15$

$\quad\quad a_{42;0} = 0.7$ $\quad a_{43;11} = 0.3$ $\quad a_{43;0} = 0.7$

Table 3.1: Aggregate statistics

Queue	Measures	SIM	MEM	%D
1	$\langle n_1 \rangle$	9.426	9.417	-0.095%
	R_1	5.032	5.025	-0.139%
	$p_1(0)$	0	0	0%
	$p_1(10)$	0.632	0.632	0%
2	$\langle n_2 \rangle$	3.854	3.920	1.713%
	R_2	2.060	2.093	1.602%
	$p_2(0)$	0.040	0.025	-37.5%
	$p_2(5)$	0.455	0.455	0%
3	$\langle n_3 \rangle$	9.283	9.344	0.657%
	R_3	4.956	4.989	0.666%
	$p_3(0)$	0	0	0%
	$p_3(10)$	0.586	0.604	3.072%
4	$\langle n_4 \rangle$	3.334	3.341	0.21%
	R_4	1.782	1.783	0.056%
	$p_4(0)$	0.070	0.068	-2.857%
	$p_4(5)$	0.314	0.313	-0.318%

Table 3.2: Marginal statistics

Queue	Class	Measures	SIM	MEM	TOL
1	1	$\nu_{1,1}$	0.202	0.174	0.028
		$\langle n_{1,1} \rangle$	1.999($\pm$0.17)	1.639	0.034
		$p_{1,1}(0)$	0.112	0.168	0.056
	2	$\nu_{1,2}$	0.358	0.304	0.054
		$\langle n_{1,2} \rangle$	3.108($\pm$0.17)	2.867	0.026
		$p_{1,2}(0)$	0.022	0.036	0.014
	3	$\nu_{1,3}$	0.440	0.521	0.081
		$\langle n_{1,3} \rangle$	4.320($\pm$0.21)	4.911	0.100
		$p_{1,3}(0)$	0.005	0.002	0.003
2	1	$\nu_{2,1}$	0.320	0.356	0.036
		$\langle n_{2,1} \rangle$	1.334($\pm$0.008)	1.431	0.009
		$p_{2,1}(0)$	0.238	0.210	0.028
	2	$\nu_{2,2}$	0.411	0.411	0
		$\langle n_{2,2} \rangle$	1.653($\pm$0.009)	1.651	0
		$p_{2,2}(0)$	0.173	0.163	0.010
	3	$\nu_{2,3}$	0.229	0.209	0.020
		$\langle n_{2,3} \rangle$	0.867($\pm$0.006)	0.838	0.005
		$p_{2,3}(0)$	0.407	0.412	0.005
3	1	$\nu_{3,1}$	0.673	0.655	0.018
		$\langle n_{3,1} \rangle$	6.244($\pm$0.07)	6.123	0.011
		$p_{3,1}(0)$	0	0	0
	2	$\nu_{3,2}$	0.287	0.309	0.022
		$\langle n_{3,2} \rangle$	2.710($\pm$0.01)	2.884	0.018
		$p_{3,2}(0)$	0.046	0.035	0.011
	3	$\nu_{3,3}$	0.04	0.036	0.004
		$\langle n_{3,3} \rangle$	0.329($\pm$0.04)	0.337	0.001
		$p_{3,3}(0)$	0.712	0.710	0.002
4	1	$\nu_{4,1}$	0.291	0.292	0.001
		$\langle n_{4,1} \rangle$	1.030($\pm$0.005)	1.049	0.002
		$p_{4,1}(0)$	0.354	0.344	0.01
	2	$\nu_{4,2}$	0.552	0.552	0
		$\langle n_{4,2} \rangle$	1.961($\pm$0.004)	1.980	0.002
		$p_{4,2}(0)$	0.154	1.148	0.006
	3	$\nu_{4,3}$	0.087	0.087	0
		$\langle n_{4,3} \rangle$	0.342($\pm$0.003)	0.313	0.005
		$p_{4,3}(0)$	0.706	0.729	0.023

5 Conclusions

A new approximate version of MEM algorithm is proposed for arbitrary multi-class open exponential QNMs with finite capacity single-server stations, unrestricted buffer sharing, RS blocking under RD and mixed service disciplines drawn from FCFS, LCFS-NPR, LCFS-PR and PS rules. A general ME solution for the censored $M/M/1/N$ queue with multi-classes of jobs is presented and used to analyse a censored queue. The later queue plays the rôle of a building block in the solution process of the entire network. Numerical examples validate the credibility of the new approximation in relation to simulation. Current work extends these results to the analysis of arbitrary open and closed networks with general interarrival and service times under various blocking mechanisms and scheduling rules.

References

[1] H.G. Perros. A Bibliography of papers on queueing networks with finite capacity queues. *Performance Evaluation*, 10:255–260, 1989.

[2] H.G. Perros. Approximation algorithms for open queueing networks with blocking. In H. Takagi, editor, *Stochastic Analysis of Computer and Communication Systems*, North-Hólland, 1990, pp 451–494.

[3] R.O. Onvural. Survey of closed queueing networks with blocking. *ACM Computing Surveys*, 22(2):83–121, 1990.

[4] I.F. Akyildiz and H. von Brand. Exact solutions for open, closed and mixed queueing networks with rejection blocking. *Theoretical Computer Science*, 64:203–219, 1989.

[5] I.F. Akyildiz and H. von Brand. Computational algorithm for networks of queues with rejection blocking. *Acta Informatica*, 26:559–576, 1989.

[6] I.F. Akyildiz and H. von Brand. Central server models with multiple job classes, state dependent routing, and rejection blocking. *IEEE Transactions on Software Engineering*, SE-15(10):1305–1312, 1989.

[7] R.O. Onvural. A Note on the product-form solutions of multi-class closed queueing networks with blocking. *Performance Evaluation*, 10:247–253, 1989.

[8] R.O. Onvural and H.G. Perros. Approximate analysis of multi-class tandem queueing networks with Coxian parameters and finite buffers. In P.J.B. King et al, editors, *Performance '90*, North-Holland, 1990, pages 131–141.

[9] D.D. Kouvatsos and N.P. Xenios. Maximum entropy analysis of general queueing networks with blocking, In H.G. Perros and T. Altiok, editors, *Queueing networks with blocking*, North-Holland, 1989, pages 281–309.

[10] D.D. Kouvatsos and S.G. Denazis. Comments on and tuning to: MEM for arbitrary queueing networks with repetitive-service blocking under random routing. Tech Report CS-18-91, Bradford University, 1991.

[11] D.D. Kouvatsos. Maximum entropy and the $G/G/1/N$ queue. *Acta Informatica*, 23:545–565, 1986.

[12] D.D. Kouvatsos and N. Tabet-Aouel. Product-form approximations for an extended class of general closed queueing networks. In P.J.B. King et al, editors, *Performance '90*, North-Holland, 1990, pages 301–315.

[13] D.D. Kouvatsos and P.Georgatsos. A maximum entropy approximation for a restricted $G/G/1/N$ queue with multi-classes of jobs under an unrestricted buffer sharing scheme. Research Report, Bradford University, 1988.

[14] M. Veran and D. Potier. QNAP-2: A portable environment for queueing network modelling. In D. Potier, editor, *Modelling Techniques and Tools for Performance Analysis*, North-Holland, 1985, pages 25–63

[15] L. Kleinrock. Time-shared systems: A theoretical treatment. *JACM*, 14(2):242–261, 1967

Performance Evaluation of FDDI by Emulation

Frank Ball and David Hutchison
Computing Department
Lancaster University
Engineering Building
Lancaster LA1 4YR
E.mail: mpg@comp.lancs.ac.uk

Abstract

This paper describes ongoing work in FDDI performance evaluation by means of an emulator. The emulator is part of an experimental system being used to investigate integration of multimedia traffic at the interface between a workstation and a multiservice network. Emulation is being used to evaluate the suitability of a number of emerging high speed networks as carriers of this type of traffic. A detailed description of one particular emulation, that of an FDDI network, is given in this paper followed by some preliminary performance results. Finally the future direction of the emulation work is outlined.

1 Introduction

The Multimedia Network Interface (MNI) project at Lancaster University is building an experimental system to study and model the integration of multimedia traffic channels at the interface between a workstation and a communications network [1,2]. A major aim of the project is to develop suitable protocols and interfaces to bridge between high-speed multi-service networks and a distributed systems platform. A key part in the MNI project is the Network Protocol Emulator (NPE) which will allow a number of the emerging high speed networks including FDDI, DQDB and B-ISDN to be assessed as the underlying network.

Using the NPE in place of a real network as a means of interconnecting the workstations has a number of attractions, cost savings being not the least of these. The standards for some of the emerging high speed networks are as yet incomplete, and implementations are unavailable. Emulation will allow the evolving draft specifications for such networks to be tested using the multimedia traffic generated by the workstations. The results from such tests may then be fed back into the standards development process. Emulation will also allow protocols which may be used to carry multimedia traffic over emerging networks to be developed in advance of network implementations.

2   Emulator requirements

Unlike a simulator, the NPE must not only model the behaviour of the emulated network, but must also provide a real time network service between the workstations. It may be considered as a black box which presents an interface to the logical link control (LLC) layer compatible with that of the emulated network, and transfers service data units (SDUs) between the corresponding LLC entities with the same performance characteristics as that network. Although the ability to offer the full performance of the emulated network is desirable, in practice it may only be possible to provide a scaled down performance when emulating certain networks.

The ultimate goal is to develop a general purpose NPE which can, by taking a specification for a network, provide an emulation for that network. Initially, however a separate emulation will be written for the different types of network. The experience gained in these initial emulations, and the studies carried out into the characteristics of individual networks, will then be used in the development of the general purpose emulator. The first of these individual emulations, that for a FDDI network, is well underway and the first iteration of the design will soon be completed.

The Fibre Distributed Data Interface (FDDI) is an ANSI standardisation proposal for a 100 Mbit/s token ring. The actual topology consists of two fibre optic rings, each capable of 100 Mbit/s and able to support 500 to 1000 stations over a maximum distance of 100 to 200 km. The stations themselves can be 2 km. apart which makes FDDI a suitable candidate for backbone networks [3,4].

The access method to the ring corresponds to the IEEE-802.5 token ring protocol, but there are several differences. The first difference is that the free token is produced by the sending station once a transmission is complete, that is once the last packet has been sent; with IEEE-802.5, a free token can only be put onto the ring when the sending station's own data packet is received back at the station. The second is that FDDI allows several packets to be sent while a station has the token; IEEE-802.5 allows only one packet to be sent. Thirdly FDDI supports two classes of data: synchronous with guaranteed bandwidth and response time, and asynchronous which uses the bandwidth unallocated to synchronous traffic. The ability to support synchronous data makes FDDI a possible candidate for the interconnection of multimedia workstations.

A target token rotation time (TTRT) is negotiated between the stations. The station requiring the shortest TTRT and hence fastest response time wins this negotiation. A station may be allocated bandwidth for the transmission of synchronous data and this will be given as a percentage of the TTRT. The total allocated to all the stations must not exceed 100%. When the token arrives at a station it may transmit frames from its synchronous queue for the duration of its allocated percentage of TTRT. The mechanism to control asynchronous access employs two timers, a Token Rotation Timer (TRT) which measures the time between the arrivals of a token at the station, and a Token Holding Timer (THT) which limits the period a station can hold the token when transmitting asynchronous data. When a station captures the token, THT is assigned the value TTRT-TRT. If this value is greater than zero (an early token) then following transmission of the stations synchronous allocation it may transmit a synchronous data until THT expires.

ANSI X3T9.5 is now looking at the next generation of high speed LANs. FDDI-II is being defined as an upward-compatible fibre-based LAN incorporating the features

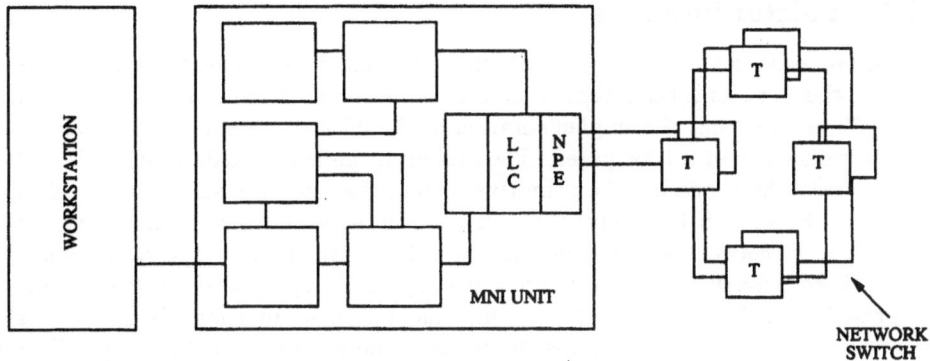

Figure 1: Emulator Configuration

of the current FDDI with an added isochronous service suitable for voice and video traffic. It must be said however that the full scope for FDDI has not been realised and that the FDDI standard is still emerging.

3 Emulator design

Three main points have had to be considered in the design of the NPE:

1. The physical interconnection of the workstations.

2. The implementation of the interface to the LLC.

3. Performance modelling of the network protocol.

We have decided to build the NPE in two distinct parts: the network switch and the protocol emulator. The network switch provides the physical interconnection and the protocol emulator provides the interface and the performance modelling.

3.1 Network switch

For the initial implementation of the network switch we have chosen to use two planes, each of four fully interconnected transputers. A packet switch running on these fully interconnected transputers transfers fixed length packets between up to four workstations. This configuration offers a bidirectional maximum theoretical bandwidth of 40 Mbit/s between any two workstations. The number of workstations that may be interconnected by this solution is limited to a maximum of four, but at some time in the future it will be replaced by some form of hardware switching mechanism that is not limited to this extent.

The initial design of the MNI workstation is implemented using transputers [2,1]. One of the workstation transputers is interconnected to the network switch by means of two transputer links, one link for each plane of the network switch. The protocol emulator along with the LLC is resident on this transputer (see Figure 1.).

3.2 Emulator interface

The issues to be considered in implementing the interface between LLC and the NPE are the same as if the interface were to be between the LLC and the network itself. We assume that the implementation of the network would also be mapped into the workspace of a transputer. The design of the interface data units (IDUs) and the methods of passing the primitives across the interface will be identical in both cases. However it is possible that not all of the services which would be offered by the type of network under emulation would be supported by the NPE. Some, such as those concerned with monitoring the physical integrity of the network, may be omitted from the emulation. In our implementation occam channels are used as the means of communicating across the interface between the LLC and the FDDI emulation, and a packet structure is used to pass the primitives and parameters.

3.3 Performance modelling

The performance modelling of the network protocol requires the design of some algorithm which will supply the appropriate timing characteristics of the network under emulation to the transfer of SDUs. The network switch will perform the transfer under control of the performance modelling algorithm. The degree to which the emulation must closely model the mechanisms of the network will depend upon the availability of reliable performance statistics for that network. Where such statistics are readily available it is possible that some analytical model may be used, otherwise a modified type of simulation may be necessary to provide the timing characteristics.

Investigations into FDDI performance have been carried out using simulation. Dykeman and Bux [5] investigate the effects of TTRT, ring latency and offered load on throughput performance. They show that the relationship between TTRT and ring latency has a major effect on throughput performance. If ring latency is large in comparison with TTRT (greater than 1/5 of TTRT) then throughput performance is degraded considerably. Johnson [6] examines various aspects of FDDI performance including average delays for both synchronous and asynchronous traffic using various ring parameter settings. The effects of offered load on the bounded token rotation time are investigated and it is shown that the upperbound of 2 * TTRT is approached only under extreme circumstances.

A major limitation to the throughput capability of the NPE is the bandwidth of the network switch. This bandwidth is less than that of many of the networks which are to be emulated. However unlike some of the networks to be emulated, the network switch can allow data to be transmitted from more than one station at any one time.

4 FDDI emulator implementation

The initial implementation of the FDDI emulator models the MAC timed token protocol fairly closely although the timing elements are much simplified. A global simulated time, measured in units of octets transmitted, is used to control station transmissions, rather than timers at each station. A token is circulated from station to station which controls transmission and carries the global timing information.

Each station copies the time carried in the token when it arrives, and updates this time before passing on the token.

By comparing the time of a token arrival with the recorded time of the token's last arrival, TRT and hence THT can be calculated. The station may then transmit frames from its synchronous queues for up to the duration of its allocated synchronous time followed by transmission from its asynchronous queues for up to the duration of THT. Once the station has completed its transmissions the total time spent in transmitting both classes of data will then be added to the time of the token's arrival. A free token will then be issued which contains the updated time. The algorithm used for the performance modelling was developed by closely examining the FDDI standards documents [7] and considering a chip set implementation of FDDI.

Due to the limited bandwidth of the transputer links the emulator can offer only scaled performance. A timer is used to compare the simulated time used by the emulator with the passing of real time. A "software probe" is inserted into the emulation code on one of the workstations. This probe monitors and reports the time of arrival of the token, the time used by the station for transmission and the amount of data transmitted by that station at that token opportunity. The logical timings are also reported for comparison.

Information from the software probe shows that the current implementation of the emulator is capable of operating at 12% of the speed of FDDI. The lower bandwidth is disguised by assuming a number of background stations which are deemed to be continually consuming 88% of the networks bandwidth. This means that only 12% of the usable synchronous bandwidth is available to be divided between the real workstations, and the minimum token rotation time will be 88% of the time available to synchronous transmission.

A major limitation to the emulator's performance is the speed of the transputer links. Extended links with differential line drivers which offer a maximum separation of 10 m are used to connect the workstations to the packet switch . The throughput of these extended links has been measured and a figure of 7.5 Mbit/s obtained for each, compared with 14.4 Mbit/s when using a standard transputer link cable. Two links are used between the workstation and the switch and therefore the upperbound to performance is 15 Mbit/s or 15% of FDDI speed. The shortfall of 3% in the measured performance of the emulator is accounted for by the processing. Current work involves the optimisation of the emulator code in order to increase performance so as to approach that determined by the transputer link speed.

5 Future directions

An optical fibre driver for extending a transputer link is being developed . This will be used to replace the differential drivers employed at present. With the optical fibre driver the physical separation between the workstation and the packet switch may be greatly increased and the throughput performance will be the same as for a link using a standard transputer link cable (14.4 Mbit/s). Once this development has been completed the emulator should be capable of performing at 25% of FDDI speed, which will become the limit of the current hardware configuration.

We are also currently considering the design of a transputer based FDDI card using a proprietary chip set supplied by AMD. This may be used to replace the FDDI emulation, requiring little or no change in the LLC implementation. It will then be possible to compare the performance of the FDDI emulator against a real implementation under identical environments. Results of this comparison may then be used to draw conclusions about the accuracy of the FDDI emulator.

The next stage of hardware development will be to replace the transputer based packet switch by some form of hardware switch. This will allow a greater number of workstations to be interconnected although initially will not offer any higher throughput since transfers between the stations will still take place via the transputer links. This will be followed by the development of a general purpose emulation engine which is intended to support the emulation of any network and offer full scale performance.

6 Conclusion

The MNI project is building a network protocol emulator which can be used to carry out experiments with multimedia traffic. The ultimate aim is to build a general purpose emulator but the initial experiments are based on a FDDI emulation. The design and implementation of the FDDI emulator have been described and future directions indicated.

Acknowledgements

The work described in this paper is supported jointly by the SERC Specially Promoted Programme in Integrated Multi-Service Communication Networks (grant number GR/F 03097) and British Telecom Laboratories.

References

[1] A. C. Scott, F. Ball, D. Hutchison and P. Lougher, "Communications Support for Multimedia Workstations", Proceedings of 3rd IEE Conference on Telecommunications (ICT'91), Edinburgh, Scotland.

[2] F. Ball, D. Hutchison, A. C. Scott and W. D. Shepherd, "A Multimedia Network Interface", Proceedings of 3rd IEEE COMSOC International Multimedia Workshop (Multimedia'90), Bordeaux, France.

[3] J. F. McCool, "FDDI: Getting to know the inside of the ring", Data Communications, 17, 3, pp 185-192 (March, 1988).

[4] F. E. Ross, "FDDI - A LAN Among MANs", ACM Computer Communication Review, 20, 3, pp 16-31 (July, 1990).

[5] D. Dykeman, W. Bux, "An Investigation of the FDDI Media-Access Control Protocol", EFOC/LAN 87, Basel, Switzerland.

[6] M. J. Johnson, "Performance Analysis of FDDI", EFOC/LAN 88, Amsterdam, Netherlands.

[7] FDDI-Part 2: "The Ring Media Access Control (MAC)", ISO 9314-2 (1989).

An Editing and Checking Tool for Stochastic Petri Nets

Auyong Lin Song

Dept. of Computer Science
University of Edinburgh

August 29, 1991

1   Introduction

Stochastic Petri nets [1] extend the modelling power of the basic Petri net model by specifying an exponentially distributed firing rate to each enabled transition for continuous time systems. This makes Petri nets useful for performance analysis.

This software tool is written as part of my MSc dissertation. The tool aims to provide a flexible and user-friendly interface for the user to create and manipulate the Stochastic Petri net model and also enables this model to be executed under an interactive simulation environment. In addition, provision is also made in the software to allow the constructed graphical model to be interfaced to the Stochastic Petri Net Package (SPNP) [2] to obtain the analytical solutions.

2   Tool Features

Three main features are incorporated in the tool, viz :

1. A graphical editor which enables the Petri net model to be built as an attributed graph.

2. An interactive simulation environment which enables the execution of the constructed graphical model under user control and allows for an estimation of some performance properties of the modelled system.

3. A graphical interface to SPNP whereby the current graph is automatically described to the package and the analytical results are made available to the user via the graphical display.

The following shall describe each of these features in turn.

2.1   Graphical Editor

The graphical editor uses the Graphical Support System (GSS) [3] as its graphical interface. It is built with the following functions:

1. Ability to create the net via activities initiated by mouse clicks and selection through menus presented in text or iconic forms. Some of its capabilities are listed as follows:

   - create a place
   - create a bounded buffer
   - create a transition
   - create a timing node
   - mark a place with a number of tokens
   - delete a node
   - create a link
   - delete a link
   - display of node and link attributes
   - enter and modify attributes for a node or link

2. Enforcement of basic syntactical rules on the created model such as

   - arcs only allowed between a place and a transition.
   - single arc in and out of a timing node

3. Provides validation check on the correct entry of the model by displaying the number of places,transitions, bounded buffers, timing nodes, arcs and floating nodes in the created graph.

2.2   Interactive Simulation Environment

The following capabilities are incorporated in the simulation environment

1. Step-by-step interactive simulation

   Tokens that are used to represent the state of the model are moved from place to place of the graphical representation of the net on firing of each enabled transition.

   The graphical display is updated on each step.

   The user can also affect the sequence of the simulation by manually selecting the transitions that should be fired first among the ones that are enabled in the current marking.

2. Setting of breakpoint

   Breakpoint can be set to match with some specified condition such as:

   - the number of tokens in a place
   - the firing frequency of a transition

3. Simulation Trace

   Produces a trace of the simulation states with respect to the advancement of simulation time.

4. Peformance Estimates

   Performance estimates are computed based on information captured during a simulation experiment. These performance estimates may be displayed on the graphical interface when required.

   It may include :

   - transition throughput (captures the idle time of transition)
   - transition firing frequency
   - lower and upper bounds for the number of tokens in a place.
   - current number of tokens in a place
   - average transit time of a token from an input place to an output place.
   - resource utilisation

5. Detection of conflict and bounded buffer violation

2.3 SPNP Interface

The graphical interface automatically constructs the net description file from the current graph and user specified input parameters.

This package produces the following analytical results in various intermediate files such as :

- reachability graph corresponding to the SPN (Stochastic petri nets)

- the continuous time markov chain derived from the SPN.

- the steady-state probability for each tangible marking.

The various output files are made available to the user via the graphical interface.

3 Tool Execution Environment

The software is implemented in the 'C' language and runs on the UNIX operating systems. The graphic interface is based on the Sunview window management system running on Sun workstations.

4 Conclusion

Due to the limited time allowed for this project, the tool is presently capable of modelling systems of average complexity and is suitable as a teaching aid on Petri Net modelling. Further extensions are possible to enhance the functionality of the tool.

References

[1] Michael K. Molloy. Performance analysis using stochastic petri nets. *IEEE Transactions on Computers,*, C-31(9), September 1982.

[2] Kishor S. Trivedi, Gianfranco Ciardo, and Jogesh K. Muppala. *Manual for the SPNP Package Version 3.0.* Duke University, Durham, NC - 27705, May 1991.

[3] Chris Uppal. *GSS4 User Guide.* The Integrated Modelling Support Environment project, March 1991.

Author Index

Abdel-Rahim, S.	88
Ball, F.	179
Candlin, R.	15
Chochia, G.	40
de C. Pinto, A.	146
Denazis, S.G.	163
Epema, D.H.J.	99
Fisk, P.	15
Georgatsos, P.H.	163
Harrison, P.G.	146
Hillston, J.	1
Hughes, P.	131
Hutchison, D.	179
Jones, M.G.W.	123
King, P.J.B.	88
Kouvatsos, D.D.	56, 108, 163
Phillips, J.	27
Pooley, R.J.	1
Skilling, N.	15, 27
Skliros, A.	56
Lin Song, A.	185
Sorensen, S.-A.	123
Tabet-Aouel, N.M.	108
Thomas, D.Ll.	73
Woodward, M.E.	150
Xenios, N.	131

Published in 1990

AI and Cognitive Science '89, Dublin City University, Eire, 14–15 September 1989
A. F. Smeaton and G. McDermott (Eds.)

Specification and Verification of Concurrent Systems, University of Stirling, Scotland, 6–8 July 1988
C. Rattray (Ed.)

Semantics for Concurrency, Proceedings of the International BCS-FACS Workshop, Sponsored by Logic for IT (S.E.R.C.), University of Leicester, UK, 23–25 July 1990
M. Z. Kwiatkowska, M. W. Shields and R. M. Thomas (Eds.)

Functional Programming, Glasgow 1989, Proceedings of the 1989 Glasgow Workshop, Fraserburgh, Scotland, 21–23 August 1989
K. Davis and J. Hughes (Eds.)

Persistent Object Systems, Proceedings of the Third International Workshop, Newcastle, Australia, 10–13 January 1989
J. Rosenberg and D. Koch (Eds.)

Z User Workshop, Oxford, 1989, Proceedings of the Fourth Annual Z User Meeting, Oxford, 15 December 1989
J. E. Nicholls (Ed.)

Formal Methods for Trustworthy Computer Systems (FM89), Halifax, Canada, 23–27 July 1989
Dan Craigen (Editor) and Karen Summerskill (Assistant Editor)

Security and Persistence, Proceedings of the International Workshop on Computer Architecture to Support Security and Persistence of Information, Bremen, West Germany, 8–11 May 1990
John Rosenberg and J. Leslie Keedy (Eds.)

Errata in *Extension to Preemption of a Method for a Feedback Queue*, D.H.J. Epema, pp. 99-107.

page 99, line 24 $\sum_{i=1}^{j} \delta_i$

page 99, line 26 $\sum_{i=1}^{j-1} \delta_i$

page 101, line 18 $G_k(t) = \frac{G(t+\Delta_{k-1})-G(\Delta_{k-1})}{G^c(\Delta_{k-1})}, \quad 0 \le t < \delta_k$

page 101, line 20 $\Delta_k = \sum_{i=1}^{k} \delta_i$

page 101, line 20 $G^c(t) = 1 - G(t)$

page 101, line 22 $p_k = G^c(\Delta_{k-1})$

page 102, line 2 $\int_0^\infty q_k(t)\,dt = q_k$

page 102, line 5 $\int_0^\infty e^{-\lambda t}\frac{(\lambda t)^n}{n!}q_{k+1}(t)\,dG_k(t)$

page 102, line 8 $\int_0^\infty e^{-\lambda t}\frac{(\lambda t)^n}{n!}(1 - q_{k+1}(t))\,dG_k(t)$

page 102, line 11 $\int_{\delta_k^-}^{\delta_k^+} e^{-\lambda t}\frac{(\lambda t)^n}{n!}\,dG_k(t) = e^{-\lambda \delta_k}\frac{(\lambda \delta_k)^n}{n!}q_{k+1}$

page 102, line 12 $e^{-\lambda \delta_k}((\lambda \delta_k)^n/n!)(1 - q_{k+1})$

page 102, line 31 $\overline{G}_k(t) = \sum_{i=1}^{k-1} p_i(1 - q_{i+1})(G_1(t) * G_2(t) * \ldots * G_i(t)) +$
                                    $p_k(G_1(t) * G_2(t) * \ldots * G_k(t)), \quad k \ge 1$

page 103, line 14 $H^*(s) = \int_0^\infty e^{-st}dH(t)$

page 103, line 15 $E[T] = -dH^*/ds(0)$

page 103, equation (1) $W^*(s) = \dfrac{s(1 - \rho)}{s - \lambda + \lambda G^*(s)}$

page 103, equation (2) $w = \dfrac{\lambda E[S^2]}{2(1 - \rho)}$

page 103, equation (3) $H_I^*(s) = H^*(s + \lambda - \lambda B^*(s))$

page 103, equation (4) $B^*(s) = G^*(s + \lambda - \lambda B^*(s))$

page 104, line 1 $\rho_k = \sum_{i=1}^{k} \lambda_i s_i$

page 104, line 8 $\overline{B}_0^*(\cdot), \overline{G}_0^{I*}(\cdot) \equiv 1$

page 104, equation (6) $R_j^*(s) = W_j^*(s + \lambda - \lambda \overline{B}_{j-1}^*(s)) \cdot \overline{G'}_{j-1}^*(s + \lambda - \lambda \overline{B}_{j-1}^*(s)) \cdot$
$$G_j^*(s + \lambda - \lambda \overline{B}_{j-1}^*(s))$$

page 104, equation (7) $W_j^*(s) = \dfrac{s(1 - \rho_j)}{s - \lambda + \lambda \overline{G}_j^*(s)}$

page 104, equation (8) $r_j = \dfrac{\lambda E[\overline{S}_j^2] + 2(1 - \rho_j)(\overline{s'}_{j-1} + s_j)}{2(1 - \rho_{j-1})(1 - \rho_j)}$

page 104, line 30 $W_j^*(s) \cdot \overline{G}_j^*(s)$

page 104, line 32 $W_j^*(s) \cdot \overline{G'}_{j-1}^*(s) \cdot G_j^*(s)$

page 105, equation (9) $w_J = \dfrac{\lambda E[\overline{S}_j^2]}{2(1 - \rho_j)}$

page 105, line 14 $E[\overline{S}_j^2]$

page 105, line 18 $G_j^*(\cdot)$

page 105, line 18 $G_j'''^*(\cdot)$

page 105, equation (11) $d_j(\delta) = \dfrac{\lambda E[\overline{S}_j^2] + 2\rho_{j-1}(1 - \rho_j)(\Delta_{j-1} + \delta)}{2(1 - \rho_{j-1})(1 - \rho_j)}, \quad j = 1, 2, \ldots$

page 105, equation (12) $w_j = \dfrac{\lambda G^c(\Delta_{j-1}) E[S_j^2]}{2(1 - \rho_{j-1})(1 - \rho_j)} + \dfrac{\lambda \int_{\Delta_{j-2}}^{\Delta_j} G^c(t) dt \cdot \sum_{i=1}^{j-1} t_i}{1 - \rho_j},$
$$j = 1, 2, \ldots$$

page 106, line 5 $\lambda \int_{\Delta_{j-2}}^{\Delta_j} G^c(t) dt = \rho_j - \rho_{j-2}$

page 106, line 5 $\sum_{i=1}^{j-1} t_i = d_{j-1}(\delta_{j-1}) + \Delta_{j-1}$

page 106, equation (14) $\dfrac{\lambda G^c(\Delta_{j-1}) E[S_j^2]}{2(1 - \rho_{j-1})(1 - \rho_j)} + \dfrac{\lambda E[\overline{S}_{j-1}^2]}{2(1 - \rho_{j-1})(1 - \rho_j)} + \dfrac{\rho_j \Delta_{j-1}}{1 - \rho_j}$

page 106, equation (15) $E[\overline{S}_j^2] - E[\overline{S}_{j-1}^2] - 2\Delta_{j-1} \int_{\Delta_{j-1}}^{\Delta_j} G^c(t) dt = G^c(\Delta_{j-1}) E[S_j^2]$